D1191269

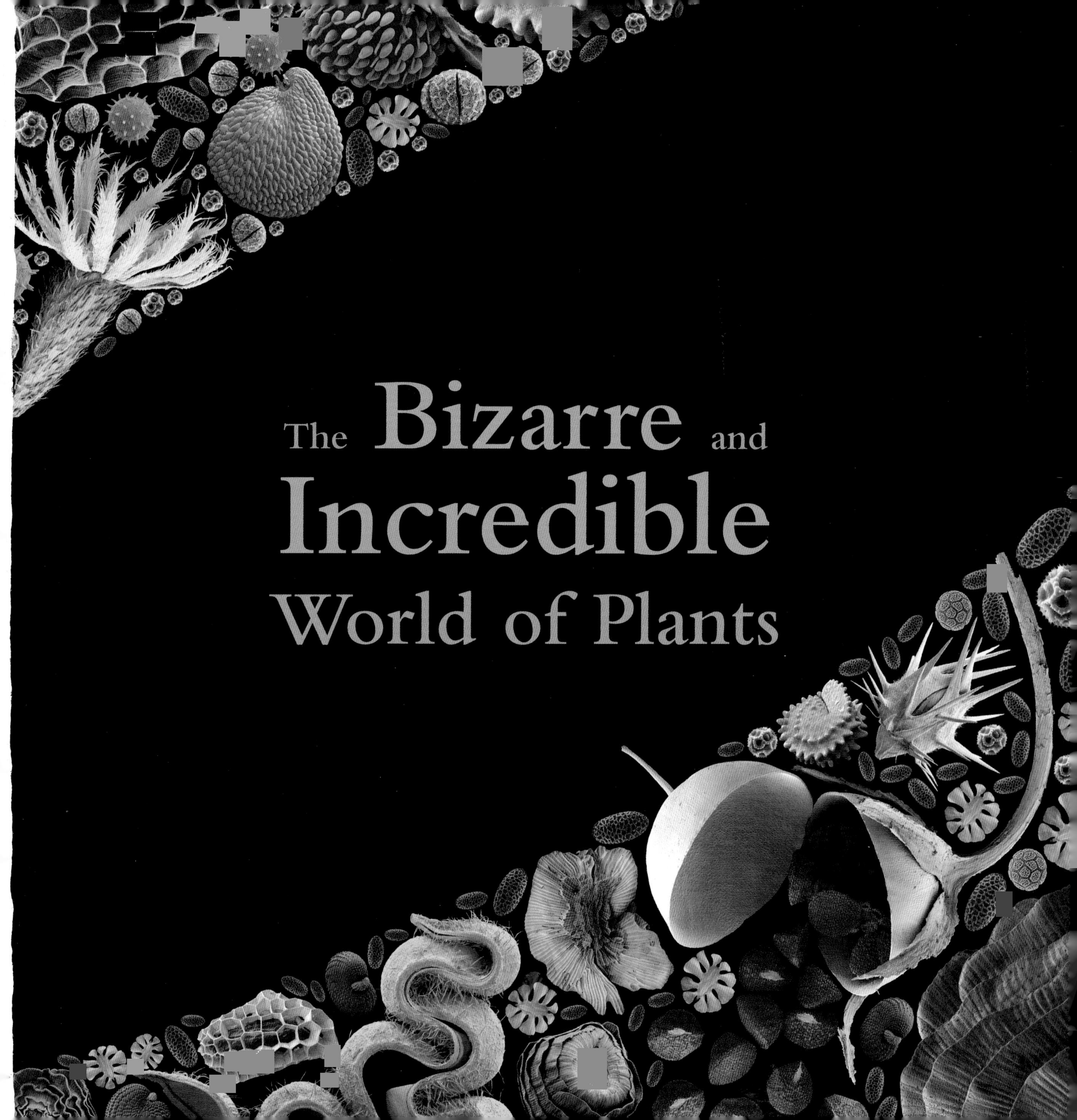

The Bizarre and Incredible World of Plants

The Bizarre and Incredible World of Plants

WOLFGANG STUPPY · ROB KESSELER · MADELINE HARLEY

Edited by Alexandra Papadakis

FIREFLY BOOKS

For Andreas, a publisher of great vision and boundless enthusiasm,

whose lively encouragement and generous support we will always remember

A FIREFLY BOOK

Published by Firefly Books Ltd. 2009

First printing

Publisher Cataloging-in-Publication Data (U.S.)

Stuppy, Wolfgang.
The bizarre and incredible world of plants / Wolfgang Stuppy.
[144] p. : 244 col. photos. ; 25 x 25 cm.
Summary: A detailed look at the world of plants, with up-close photos
showing gorgeous details from a variety of seeds, pollen, fruit and flowers.
ISBN-13: 978-1-55407-533-1
ISBN-10: 1-55407-533-5
1. Plants -- Miscellanea. I. Title.
581 dc22 QK50.S787 2009

Library and Archives Canada Cataloguing in Publication

Stuppy, Wolfgang
The bizarre and incredible world of plants / Wolfgang Stuppy, Rob Kesseler, Madeline Harley.
Includes index.
ISBN-13: 978-1-55407-533-1
ISBN-10: 1-55407-533-5
1. Plants. 2. Plants--Pictorial works.
I. Kesseler, Rob II. Harley, Madeline
III. Title.
QK45.2.S79 2009 580 C2009-903183-3

Published in the United States by
Firefly Books (U.S.) Inc.
P.O. Box 1338, Ellicott Station
Buffalo, New York 14205

Published in Great Britain by
Papadakis Publisher
A member of New Architecture Group Ltd.
Kimber Studio, Winterbourne, Berkshire,
RG20 8AN, U.K. www.papadakis.net

Published in Canada by
Firefly Books Ltd.
66 Leek Crescent
Richmond Hill, Ontario L4B 1H1

Editorial and Design Director: Alexandra Papadakis
Editor: Sheila de Vallée
Editorial Assistant: Sarah Roberts
Intern: Naomi Doerge

Printed and bound in China

CONTENTS

INTRODUCTION

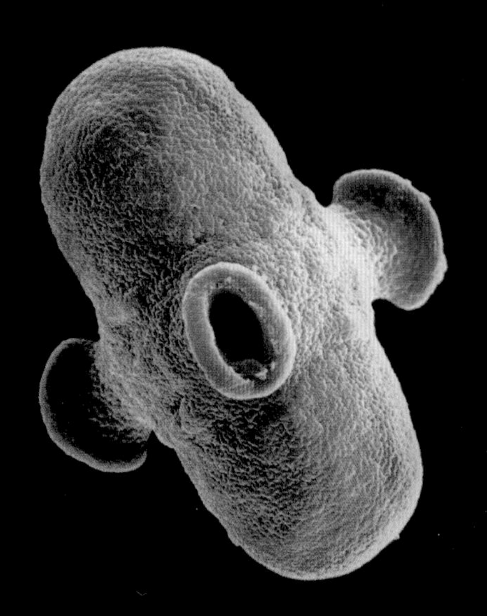

Plants first conquered the Earth's landmasses an unimaginable six hundred million years ago. From the early spore-bearing plants that were similar to our mosses and ferns, it would be another two hundred and forty million years before the evolution of pollen and seeds, two of the most crucial innovations in the history of all life on our planet. Seed plants have continued to evolve for the past 360 million years. Many remarkable adaptations for ensuring their survival, which involve their flowers, pollen, seeds and fruits, have occurred, and their methods of sexual reproduction have been perfected. Previously, in a series of three books, we explored the science, natural history and aesthetic beauty of the private life of plants. The combined expertise of an artist (Rob Kesseler) and two scientists (Madeline Harley and Wolfgang Stuppy) led to *Pollen– The Hidden Sexuality of Flowers*, *Seeds – Time Capsules of Life,* and *Fruit – Edible, Inedible, Incredible*. *The Bizarre and Incredible World of Plants* draws on images of some of the most spectacularly hidden (without light and electron microscopy) but vital aspects in the life of plants hitherto largely unknown outside the scientific community.

Art and Science

Prior to the second half of the 20th century there was a vital, shared passion and appreciation of the plant world in which thoughtful observers all played a part, including artists and artisans. Many serious amateurs contributed to, or challenged the work of early botanical scientists, who were frequently what we would now refer to as 'polymaths', 'scientists' being a rather cold, later 20th century term which has caused in people's minds an unrealistic rift between general powers of observation and thought, and how we later illustrate and record our observations.

Realising the untapped potential for working with scientific data from microscopic plant material, and to foster new audiences for the important work done at the Royal Botanic Gardens, Kew, we joined forces to bridge the gulf that had gradually developed between art and science.

Colour-coded messages

Colour in nature, in science and in art fulfils many different functions. Plants have evolved a sophisticated spectrum of colour-coded messages to attract animals to ensure the pollination of their flowers and the dispersal of their seeds. The scientist uses colour to facilitate discussions within the botanical scientific research community. Here, the artist, Rob Kesseler has used colour to enhance the beauty and expressiveness of high-magnification black-and-white scanning electron micrographs of pollen and seeds to engage a wider public audience. His choice of colours is a personal one and may relate to the original plant or be used to reveal functional characteristics of the specimen. It is used intuitively to create mesmerising images that lie somewhere between science and symbolism, sensual markers inviting further contact with unseen miracles of the natural world.

Our work results from an enthusiasm to communicate and express our shared passion and fascination for the beauty of sexual reproduction among flowering plants, which is revealed in all its splendour using methods and technologies not available until the late 20th century. We hope that we will continue to engage new audiences for the important research and conservation work in the plant sciences being carried out throughout the world. This includes the Royal Botanic Gardens, Kew and particularly the Millennium Seed Bank Project (MSBP), one of the largest international conservation initiatives in the world; much of the material illustrated in our books came from the collections of the MSBP.

Wolfgang Stuppy, Royal Botanic Gardens Kew – Wakehurst Place, Millennium Seed Bank

Rob Kesseler, Central Saint Martins College of Art and Design, London

Madeline Harley, Micromorphology Unit, Jodrell Laboratory, Royal Botanic Gardens Kew

June 2009

THE INCREDIBLE LIFE OF PLANTS

Plants are truly amazing because, unlike animals, they have the remarkable ability to use sunlight to make sugar from just water and carbon dioxide (*photosynthesis*). In doing so, they not only produce their own food but also feed – either directly or indirectly – all life on Earth. Furthermore, as a by-product of photosynthesis, they produce the oxygen in our atmosphere. Quite simply, without plants we would not be able to breathe or eat. Rice alone is the staple food of over half of the Earth's population; and there are many other cereals, as well as pulses and vegetables. Apart from essential nourishment, plants give us delicious treats such as fruits, nuts and precious spices, and useful things like timber, fibres and oils.

Plants play an important role in our lives in many different ways but, because they are static and silent, we tend not to consider them as living entities like ourselves. Their completely different texture and appearance, and the fact that they are rooted in the ground and move on a time scale that is far too slow to be noticeable to the human eye seem to render any comparison with animals and humans absurd, but this is far from true. Not only do plants have lives, just as animals do but, over several hundred million years of evolution, like animals, they have developed very complex lives, often in mutual response to animal evolution. Despite their differences, plants and animals share the same purpose in life: survival to achieve sexual reproduction and ensure the continuity of the species. However, plants, unlike animals, have a back up strategy: in the event of unrequited love they can often reproduce asexually. Nevertheless, sexual reproduction is critically important, and this is why: a new animal begins life as the result of the union of a sperm from the father and an egg cell from the mother. In the process, each parent contributes one set of *chromosomes*. The same happens in plants when a male sperm and a female egg meet. In all living beings, the chromosomes contain the genes that determine every characteristic of the organism. By mixing together the chromosomes and thus the genetic traits of the parents, an offspring with a slightly different, perhaps even better combination of characteristics is created. Furthermore it is sexual reproduction that provides the basis for evolution

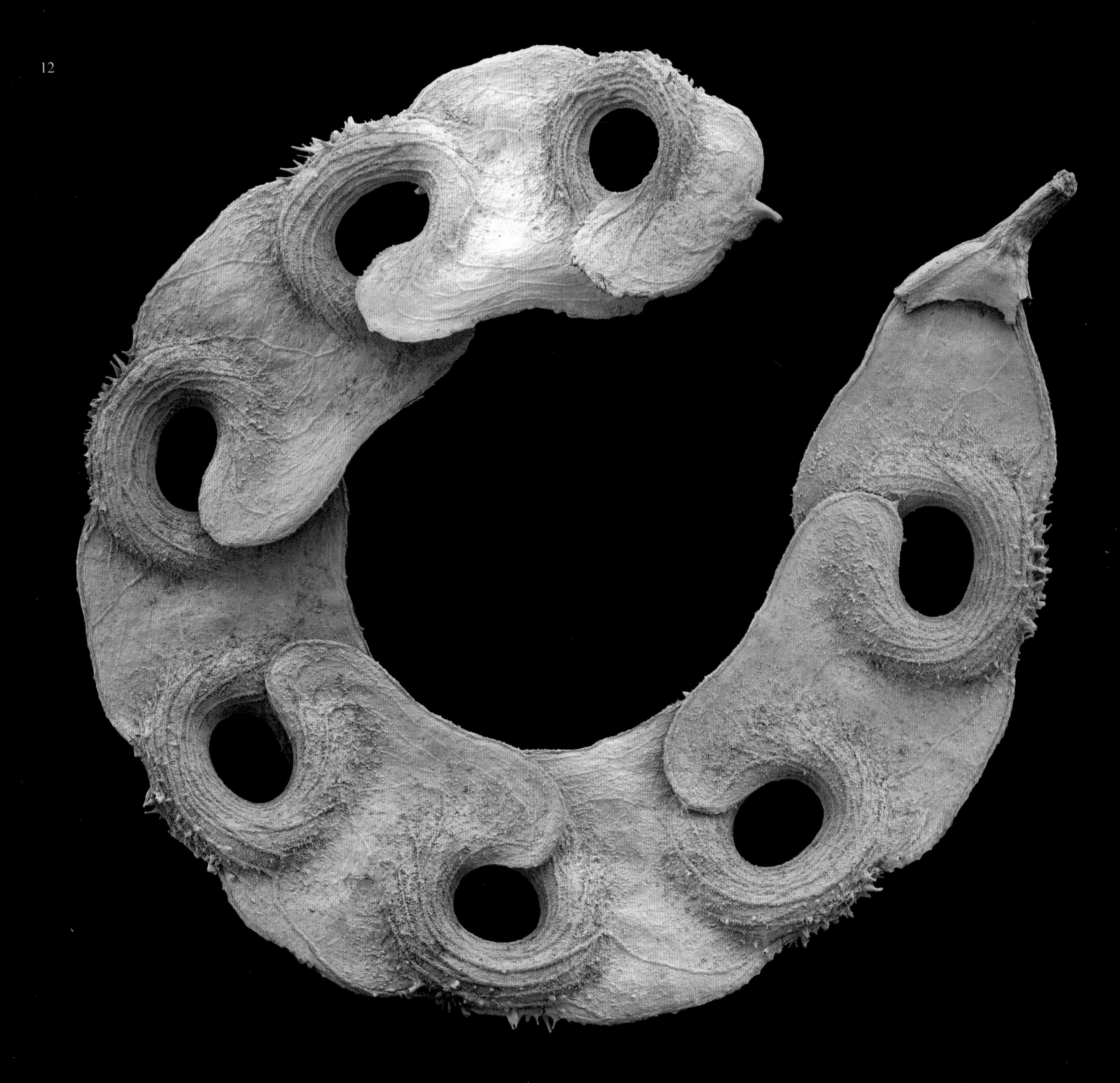

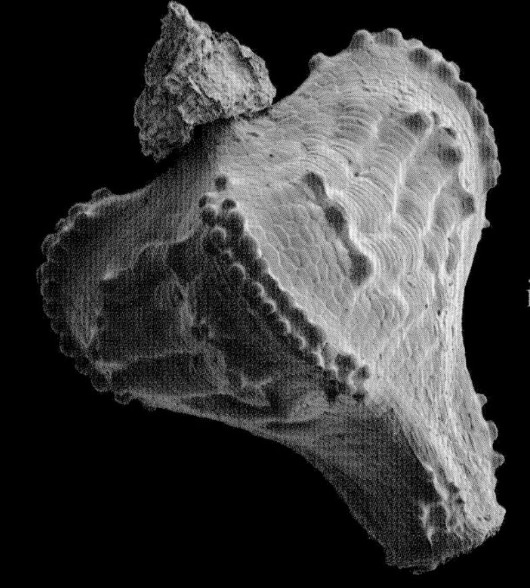

by natural selection. Many plants can reproduce vegetatively, for example the runners of strawberries, but the new individuals are genetically identical clones of the mother plant and this is why most plants typically reproduce sexually. That plants have a sex life still comes as a surprise to many people although we are familiar with the activities that surround their sexual activity. Perhaps without realising what is going on, we enjoy watching some of the ways in which plants conduct their most private affairs: flowers are pleasing to the eye and and often to the nose as well, and the fruits that follow bring pleasure to our palate.

However, from a scientific point of view, flowers are simply a display of often colourful, insect-attracting petals surrounding the central male and female genitalia – the *stamens* and the *pistil*. After sexual union, as the flowers fade, fruits develop from the female ovaries at the base of the pistil. Fruits are swollen female organs which carry the tiny plant embryos, each packaged within a *seed coat*. After the seed ripens and leaves the parent plant the embryo inside the seed coat will germinate and, leaving the safety of the seed coat, develops into a seedling that will give rise to a new plant carrying the full chromosome complement from both parents.

Floral sexual organs, and the fruits and seeds which develop after sexual union bear an enormous responsibility: flowering, pollination and fruiting are the key events in a plant's life and vital to the survival of the species. It is because of the union of sperm – carried by pollen grains – with ovaries of a plant that fruits develop and carry seeds which are the next generation of plants. It is hardly surprising, therefore, that plants have evolved a huge variety of strategies to ensure the success of their progeny.

PRECIOUS DUST

Sexual reproduction in plants is basically the same as in animals (and that includes us). For a plant to reproduce sexually, a sperm cell has to fertilise an egg cell to create the next generation; to achieve this the sperm always plays the active role in finding an egg cell to fertilise. However, because, unlike most animals, plants are unable to move about to find a mate of the same species, they have developed some very clever strategies, frequently involving animals – mainly insects – in order to achieve their objective of getting a sperm to meet with an egg cell. How on earth do they do this we might ask. The answer lies in the flower, a hot bed of sexuality, where the male and female reproductive organs are contained.

A typical flower consists of four or five whorls of highly specialised parts. The outermost whorl, the *calyx*, is a cup-shaped structure of, usually, small green leaves, the *sepals*. Within the calyx is the usually larger and often brightly coloured *corolla*, commonly made up of three to five *petals*. Between or opposite the petals, one or two whorls of *stamens* are inserted. These are the male organs of the plant. The stamens encircle, in the centre of the flower, the female organ – the *gynoecium*, or in more general, but less scientifically correct usage, the *pistil*.

The gynoecium consists of one or several *carpels*, basically modified fertile leaves folded along their midrib and fused along their opposite margins to form a bag that contains the immature seeds (*ovules*). The carpels may be separate, as in Buttercups (*Ranunculus* species, Ranunculaceae), or united into a single pistil, as in an orange (*Citrus* x *sinensis*, Rutaceae) where each fruit segment represents one carpel.

A stamen consists of a slender stalk, the *filament*, carrying the *anther* at the top. The anther, which usually has four pollen sacs (*locules*), is the fertile part of the stamen and produces thousands of microscopically small dust-like grains called *pollen*. Each pollen grain carries a tiny but precious cargo of two male sperm. To deliver their sperm, the pollen grains must somehow reach the female organs of the same flower or, preferably, of a flower on another individual of the same species. The gynoecium (or pistil) is divided into the *ovary*, the swollen fertile part at the base, and the

stigma, a special pollen-receiving area on top of the ovary. Sometimes the stigma is raised above the ovary by a column-like extension (the *style*). Pollen grains landing on the wet surface of the stigma rehydrate and germinate within minutes, sending out a tube-like extension. This *pollen tube* penetrates the surface of the stigma and grows downwards through the style until it reaches the ovary, which is literally the womb of the flower and contains one or many tiny immature seeds (*ovules*), each bearing a single egg cell. To fertilise an egg cell, a pollen tube must enter the ovule through a tiny opening in the female tissue, called the *micropyle*. Once inside the ovule, the tip of the pollen tube ruptures and releases two sperm, one of which fertilises the egg cell while the other fuses with the *polar nuclei* of the ovule to form the *primary endosperm nucleus* from which the seed storage tissue, the *endosperm*, will develop. Whereas animals and humans have motile sperm, plant sperm have to be delivered directly to the egg cell by the pollen tube. Once fertilised the egg cell develops into a baby plant, the *embryo*, and the ovule becomes a seed.

This is how seed-bearing plants – flowering plants (*angiosperms*) and conifers and their allies (*gymnosperms*) – go about their sex lives. However, mosses, liverworts, ferns and their allies reproduce via spores rather than seeds and the life cycle of these plants has some notable differences.

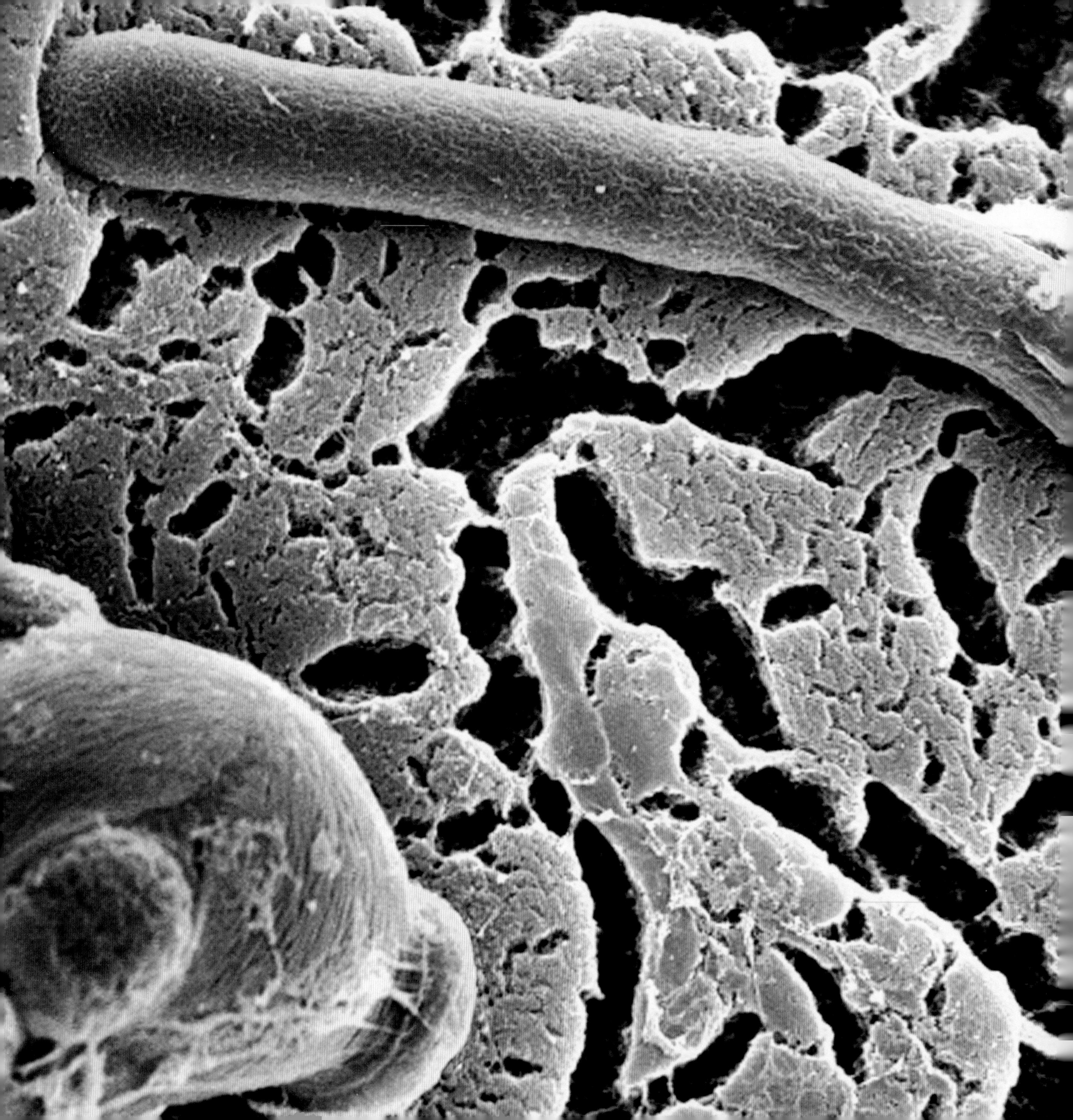

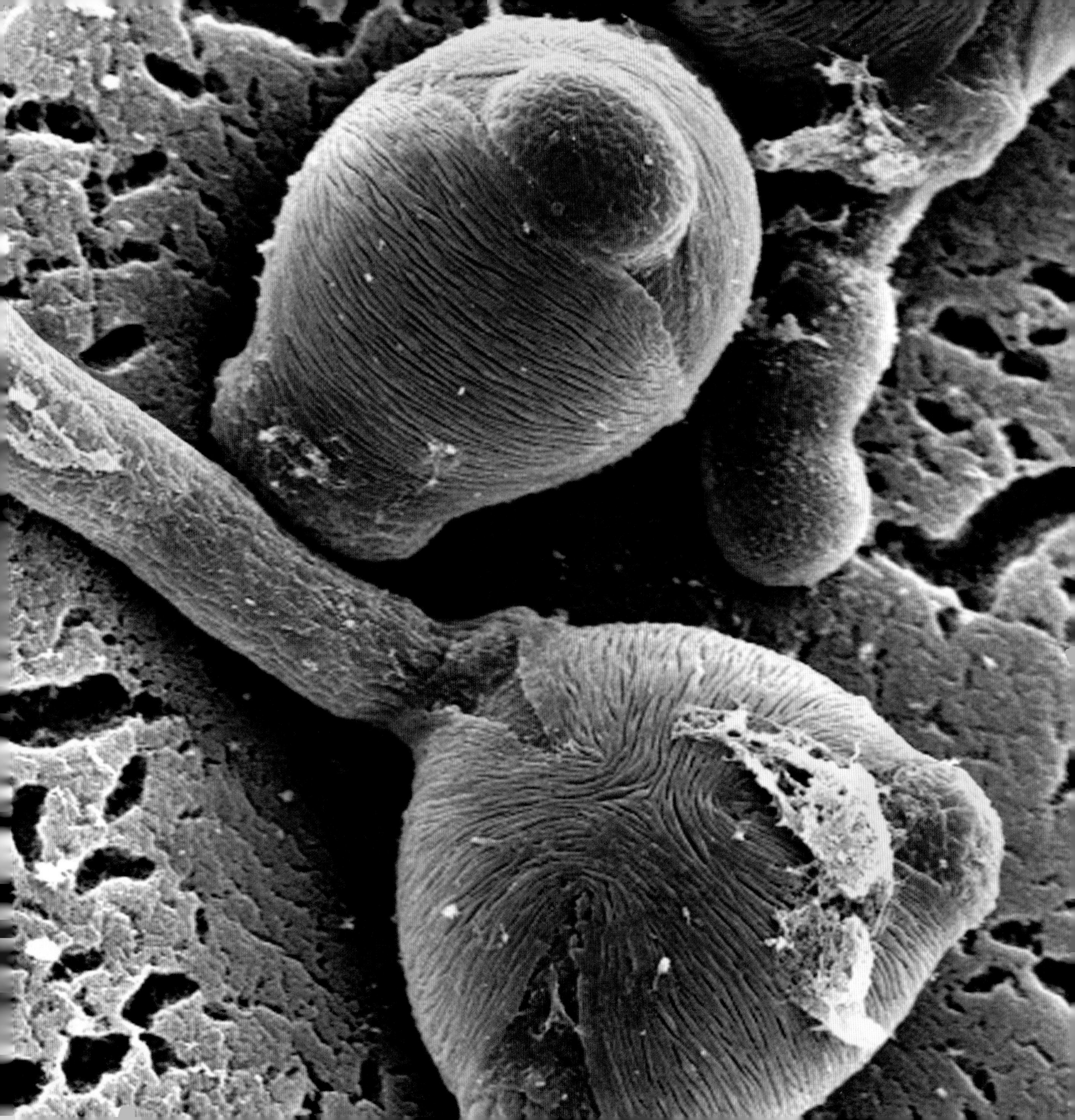

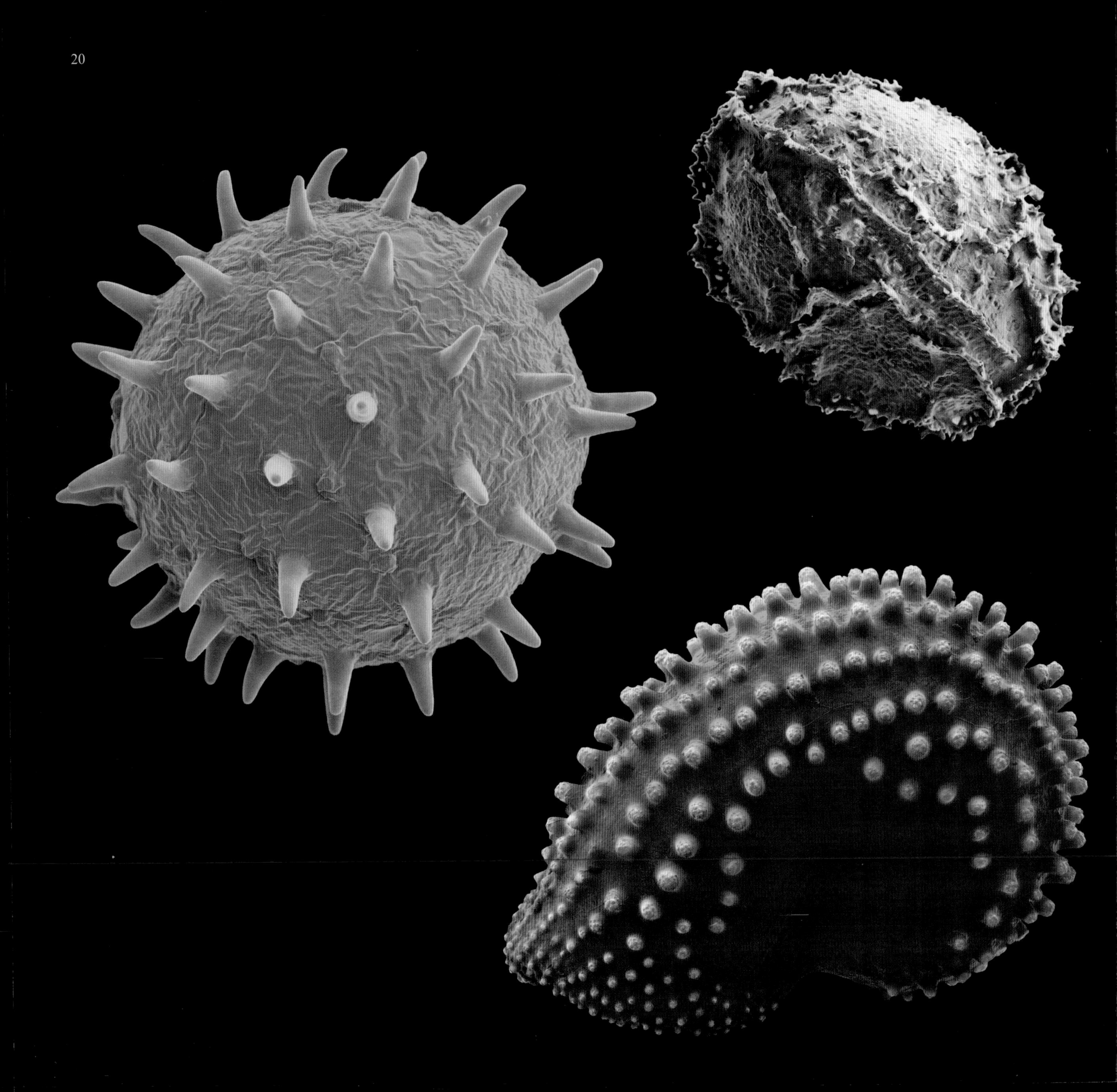

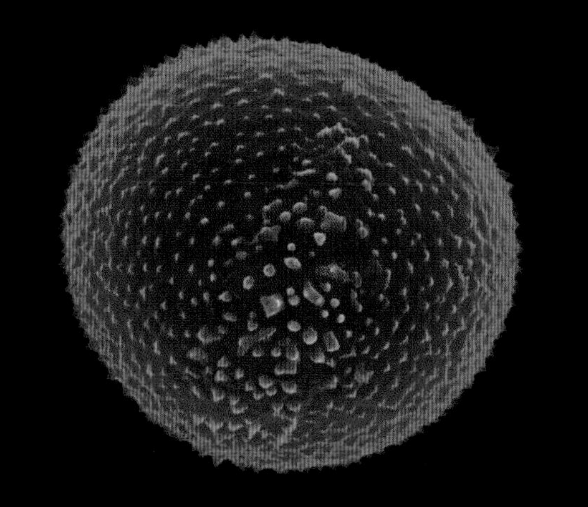

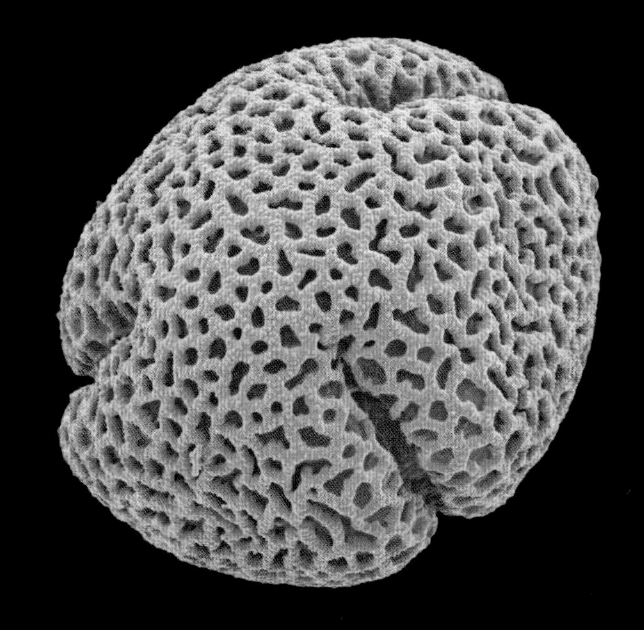

THE DIFFERENCE BETWEEN POLLEN, SPORES AND SEEDS

Since they both originate from plants and both have a dust-like appearance spores are often likened to pollen. However, there is a fundamental difference between pollen grains and spores. *Alternation of generations* (those with haploid chromosomes and those with diploid chromosomes) is unique to plants, from green algae, through liverworts, mosses, ferns, conifers and their relatives (gymnosperms) to flowering plants. It has no counterpart in the animal kingdom. In principle all plants share the same life cycle, but there is a big difference between seed-bearing plants (gymnosperms and angiosperms – the *spermatophytes*) and spore-bearing plants (the *cryptogams*, a choice of name that proves that some scientists have a sense of humour: it means 'those who copulate in secret').

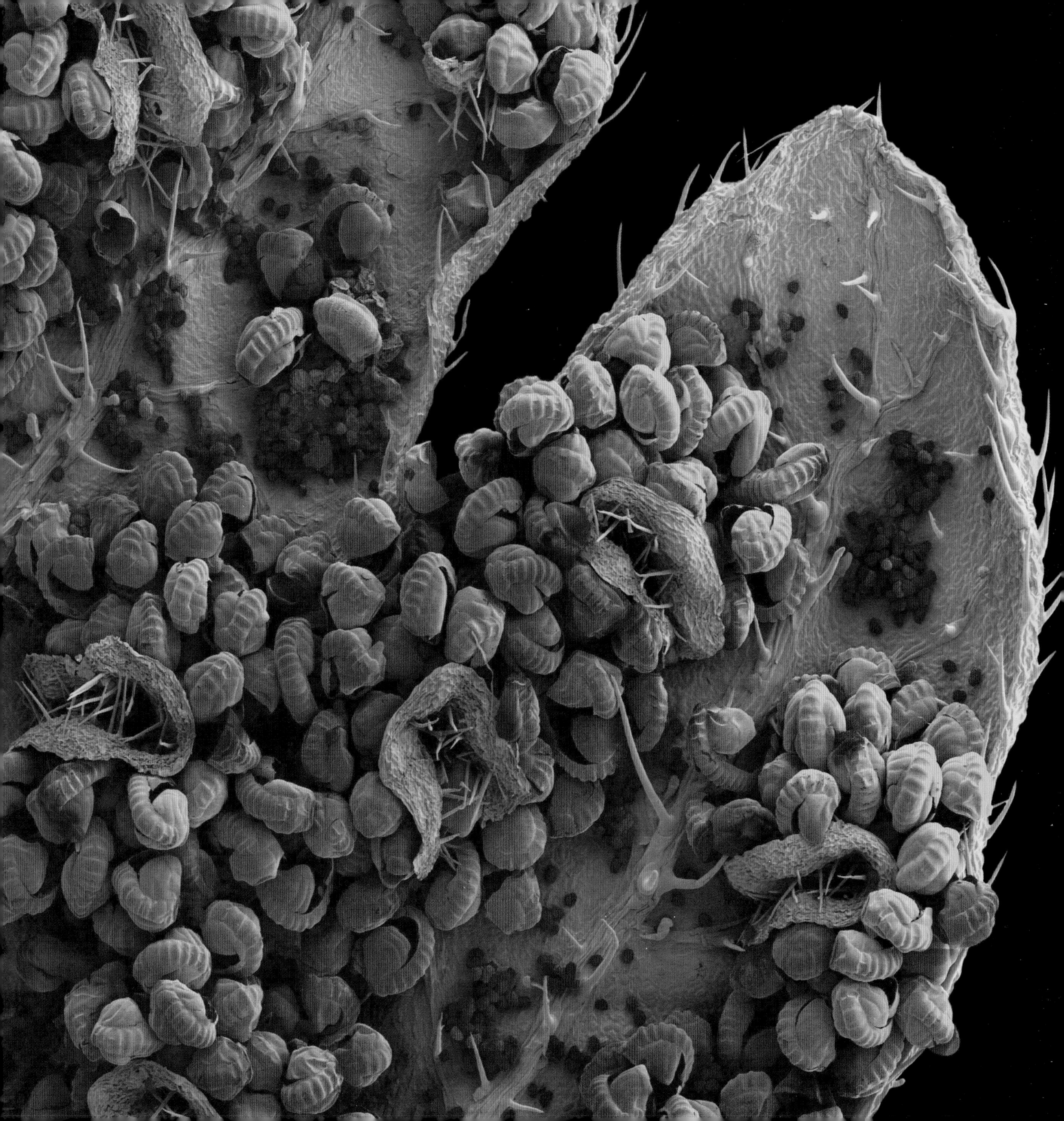

Unlike angiosperms and gymnosperms, cryptogams also have an independent, often photosynthesizing (green) haploid generation, the *gametophyte.* To give a well known example, look on the underside of a mature fern frond – such as the male fern *Dryopteris filix-mas*; on the individual leaflets (*pinnae*) there are rows of tiny kidney-shaped structures, each one is a *sorus* (plural *sori*). Protected under each sorus are the sporangia, which contain the spores. When sporangia mature they burst open to release the spores, which, like pollen grains, are the haploid generation. In moist conditions on the ground the spores germinate and grow into tiny haploid gametophytes. These gametophytes appear very different from the fern plants we are familiar with; many look more like a liverwort than a fern. On reaching maturity, the gametophytes produce male organs (*antheridia*) on the underside of their fronds; these release motile sperm and female organs (*archegonia*) containing egg cells. In the presence of water (for example, rain, dew, spray from a river or waterfall), the sperm cells are released from the antheridia of one gametophyte and swim across to the egg cells waiting for them in the archegonia on another gametophyte. Just like the gametophytes that produced them, the sperm and the egg cell are both haploid and contain only one set of chromosomes.

After fertilization the egg cell will contain two sets of chromosomes – it has become diploid – and is called a *zygote*. The zygote then develops into the impressive and often beautiful plant that we recognize as a fern. A new generation of haploid spores will be produced inside the sporangia on the underside of the fronds of this diploid fern plant, which is why this generation is called the *sporophyte* (literally *spore-producing plant*).

The biggest handicap of cryptogams is that they have motile sperm that require water to swim across to an egg cell in order to achieve fertilization. This is a real disadvantage when living on land, and today's spore-producing land plants, such as mosses, clubmosses, horsetails and ferns, still have not solved this problem internally and have to rely on external support, which is why they are usually found growing in humid environments or at least in areas where wet periods are common events interspersed in the otherwise dry conditions. This explains the presence of xeric ferns – such as Lip Ferns (*Cheilanthes* species) and the False Rose of Jericho (*Selaginella lepidophylla*), a clubmoss, in semi-arid habitats such as the North American deserts.

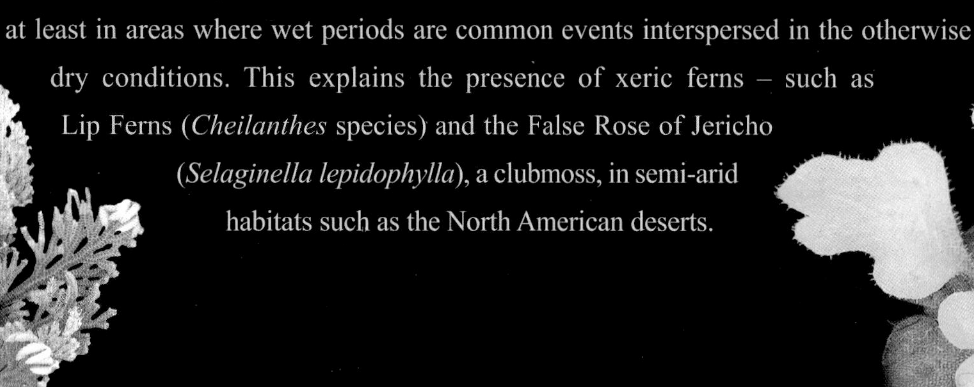

NO MATCH FOR A SEED

The appearance of pollen and spores is similar. In fact, pollen grains are male spores that have lost the ability to germinate on the ground and grow into independent gametophytes. For a pollen grain to produce a pollen tube (the reduced male gametophyte of a seed plant), it needs the nurturing substrate of a stigma (or *pollen chamber* in gymnosperms). Although pollen grains may appear very similar to spores, there is nothing in the life cycle of a spore plant that corresponds to a seed. Unlike spores, which produce the haploid generation, when seeds germinate they produce the diploid generation (sporophyte).

The evolution of pollen, ovules and seeds, which is unique to gymnosperms and flowering plants, was the fundamental step in the evolution of land plants. It won them independence from water for sexual reproduction, an enormous advantage over spore-bearing plants. In seed plants, the fertilised egg cell develops into a new sporophyte (the embryo) within the safety of the ovule. Unlike cryptogams, where the zygote must grow into a new sporophyte immediately, the embryo of a seed plant grows only to a certain size and then often waits inside the seed (a mature ovule) for optimal germination conditions. The temporarily inactive embryo has a food reserve (endosperm) provided by the mother plant, and is protected from desiccation and damage by the seed coat. The evolution of the seed is significant among vascular plants and can be compared to the evolution of the shelled egg in reptiles. Just as the seed allowed plants to escape their dependency on moist habitats, the egg enabled reptiles to become the first fully terrestrial vertebrates. In that sense, mosses, liverworts, ferns and fern-allies are more like amphibians which still rely on water for fertilization despite their terrestrial existence.

Later we will show just how extraordinary seeds are, but before this we should take a closer look at pollen, which is also quite remarkable.

AN INVISIBLE MICROCOSM

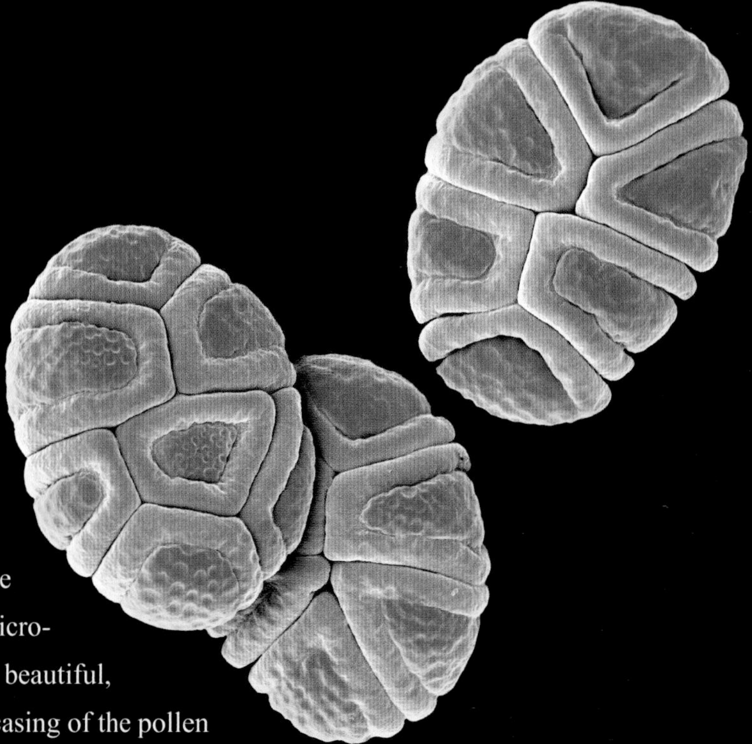

Most of us are aware of pollen, mainly because it can stain our clothes or, more annoyingly, cause allergic reactions in the form of hay fever. Leaving these irritants aside, upon close examination, pollen grains reveal themselves as perfect masterpieces of natural architecture and structural engineering. With an average size of between 20-80 *microns* (a micron is one thousandth of a millimetre) most pollen grains are tiny to the point of being invisible to the naked eye. Nevertheless, many are breathtakingly beautiful. If we look at pollen grains down a microscope we enter a fantastic microcosm where, although small is beautiful, use far outweighs ornament. The appearance of the tough outer casing of the pollen grain, which encloses the sperm cells, shows an amazing range of variation between different species of plants. These variations are frequently elaborate, often exquisitely so, and they are referred to as 'pollen types'. There are thousands of pollen types. Usually a plant species produces pollen of only one type. However, there are not as many pollen types as there are species of plants, and some species share a very similar pollen type with another species, particularly species which are closely related. Some pollen types are common to a number of plant families, and if the plant that produced the pollen is not to hand, it is difficult, even for an expert, to identify the plant that produced the pollen. Then there are families of plants, such as the grasses (Poaceae), where the pollen is extraordinarily similar in all the species but, nevertheless, highly recognisable as grass pollen.

In most plants pollen is released from the anthers of mature flowers as individual grains. However, in about fifty plant families there are at least some species where the mature pollen grains are dispersed in groups of four, known as tetrads. These include many members of the heath family (Ericaceae), as well as the Evening Primrose family (Onagraceae) such as Fuchsia and Rosebay Willowherb (*Epilobium angustifolium*). Pollen may also be shed in larger groups, called polyads. In polyads the grains are normally in multiples of four. Polyads occur, for example, in species of *Acacia* and *Mimosa* (Leguminosae, subfamily Mimosoideae). Another pollen dispersal unit, the pollinium occurs in two other very large families, the orchids (Orchidaceae) and the Milkweed family (Asclepiadaceae; now treated as a subfamily of the dogbane family, Apocynaceae). Here the pollen grains are exposed in more or less compact, coherent masses (massulae).

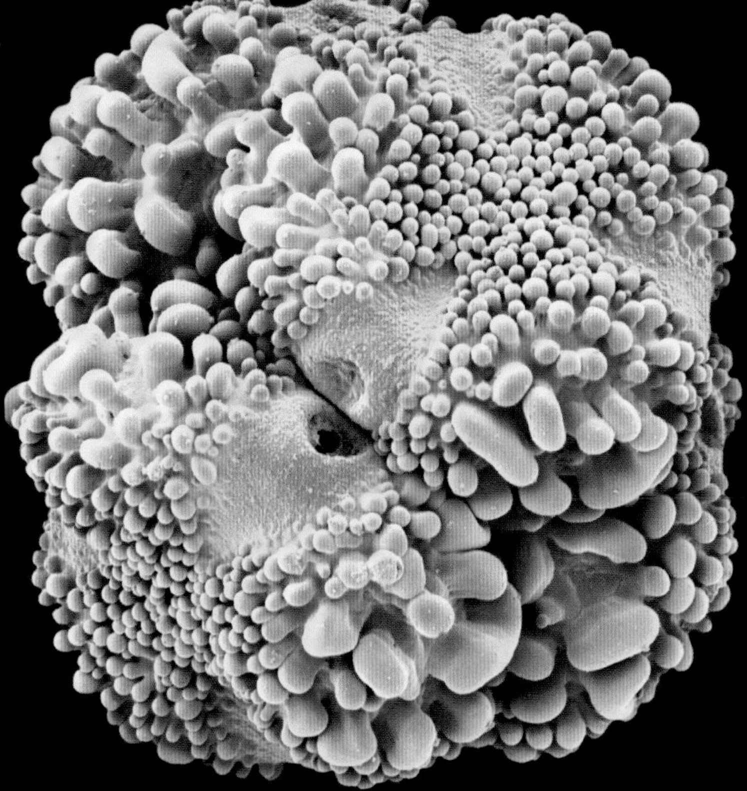

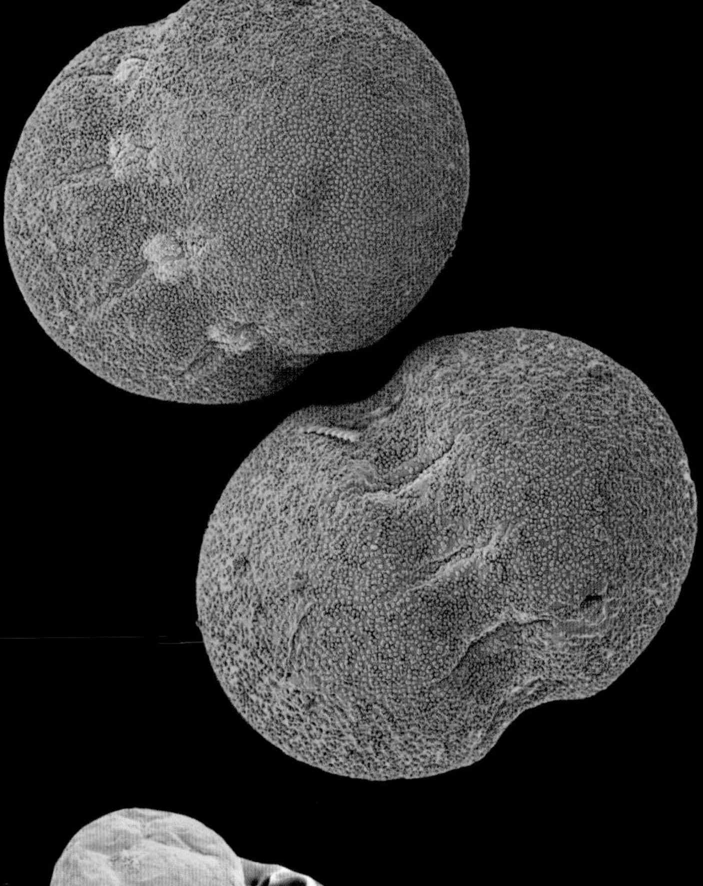

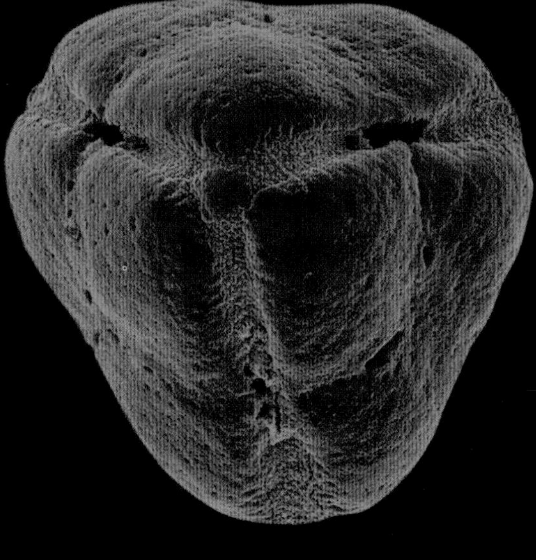

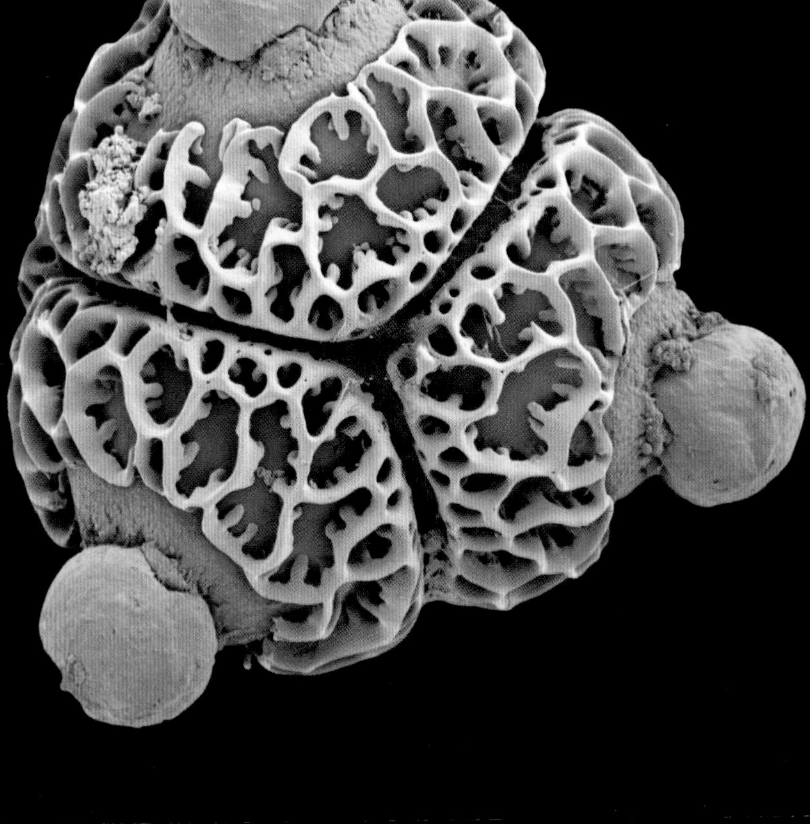

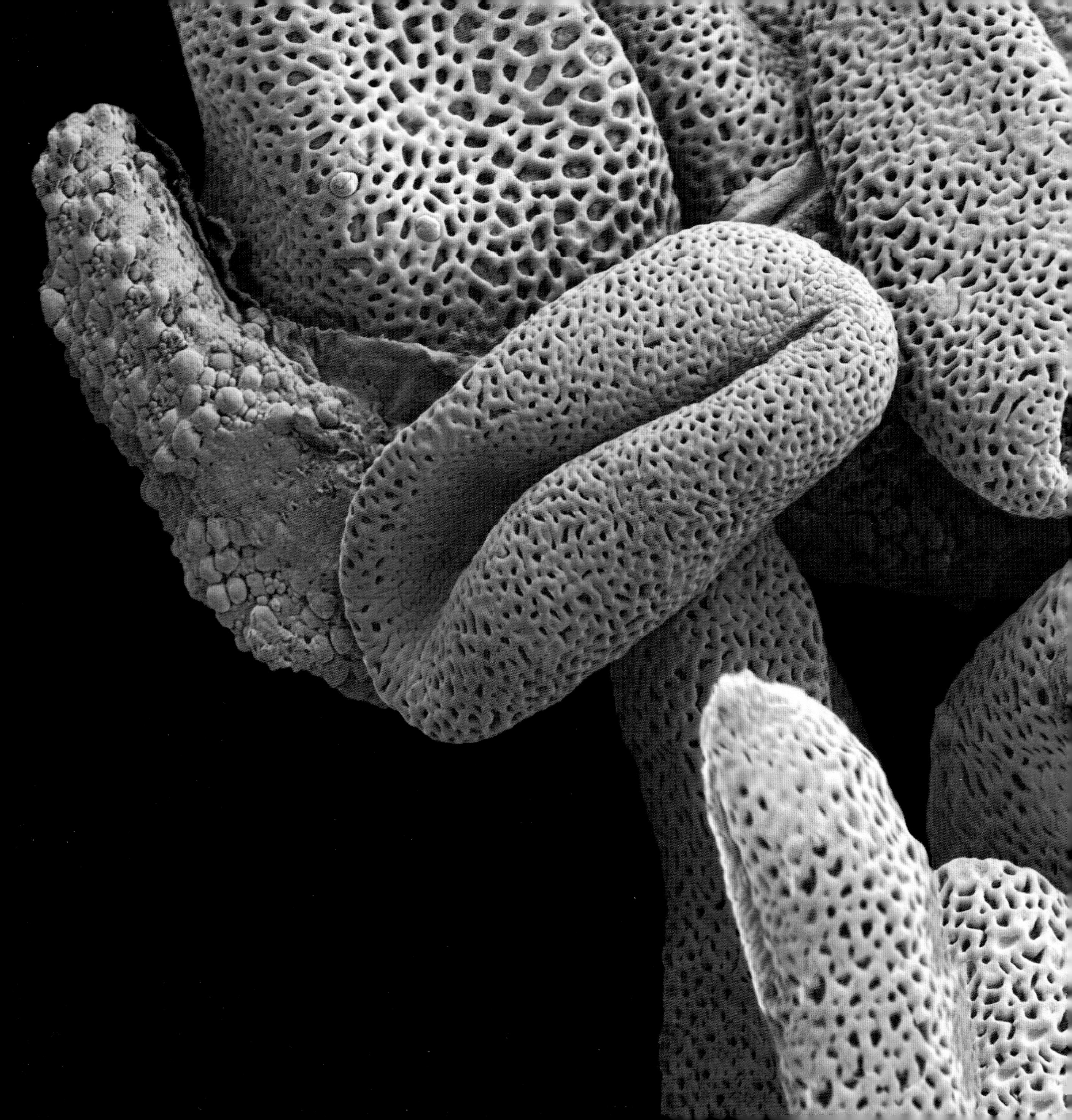

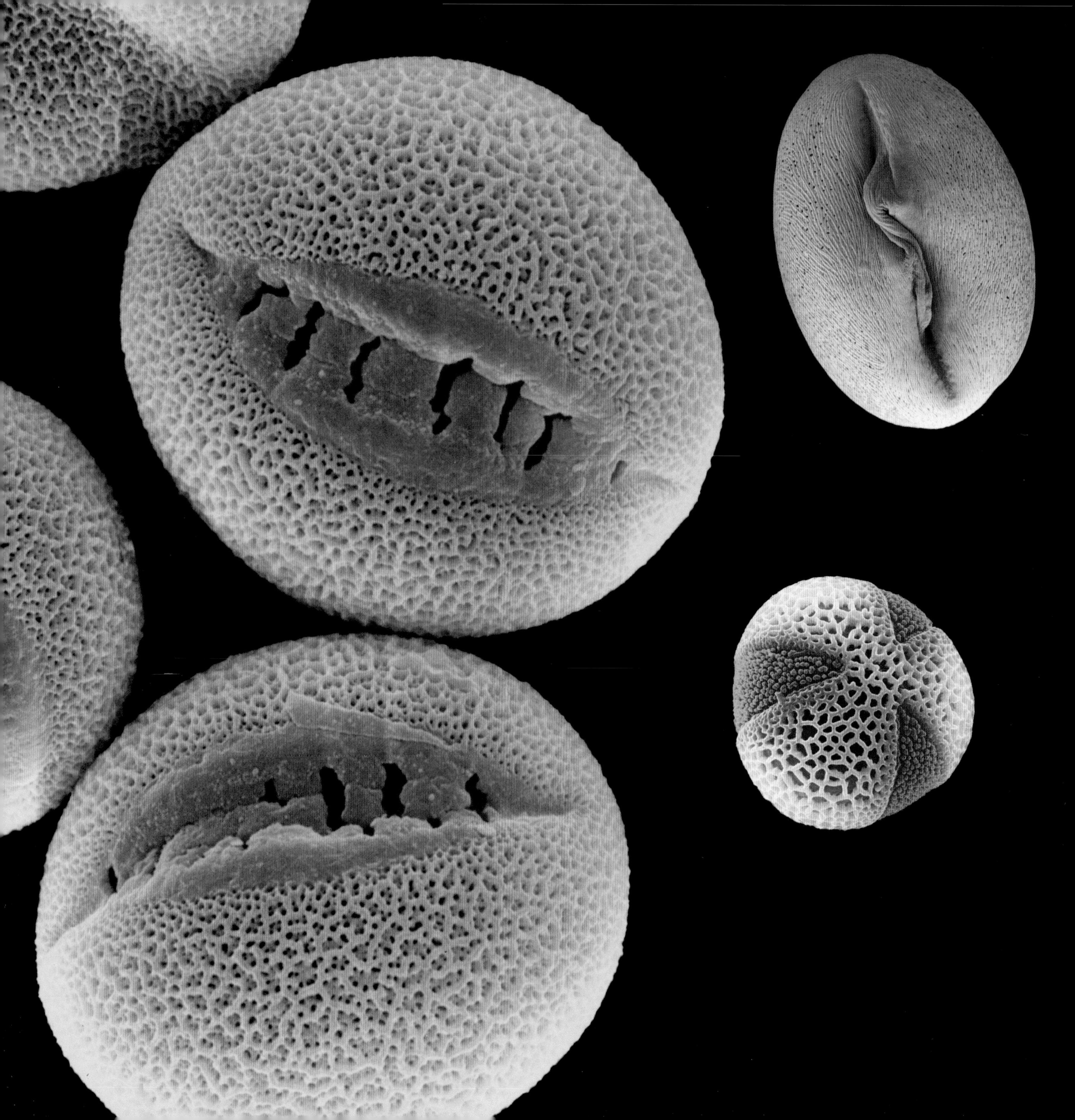

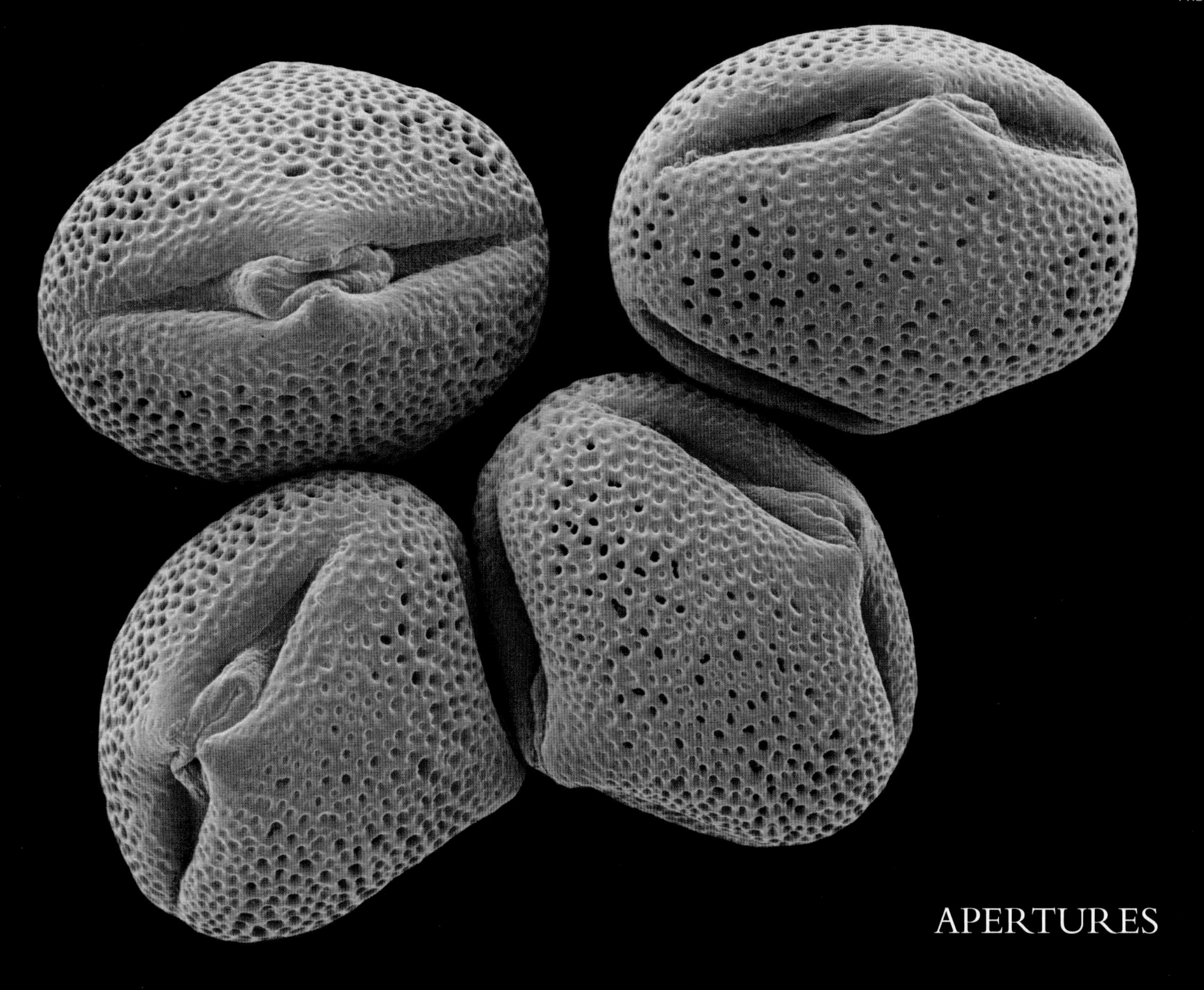

APERTURES

Apertures are important functional characteristics of most pollen grains. They are specialised openings in the pollen wall through which the germinating pollen tube carrying the sperm cells from the pollen to the ovule will exit. The number of apertures in a pollen grain varies from one to many among different species of plant. The earliest fossil pollen grains recovered, dating to about 120 million years ago, have just one elongate, slit-like aperture. The pollen of Magnolias (family Magnoliaceae) and Palms (family Arecaceae) still carries this characteristic and both groups of plants represent very early-evolving angiosperm families. However, pollen grains with three simple, radially distributed, elongate apertures are also found very early in the pollen fossil record, and this aperture arrangement, too, is still found in many living plants such as the Christmas Rose (*Helleborus niger*, Ranunculaceae), Witch Hazel (*Hamamelis* species, Hamamelidaceae) and Maples (*Acer* species, Sapindaceae).

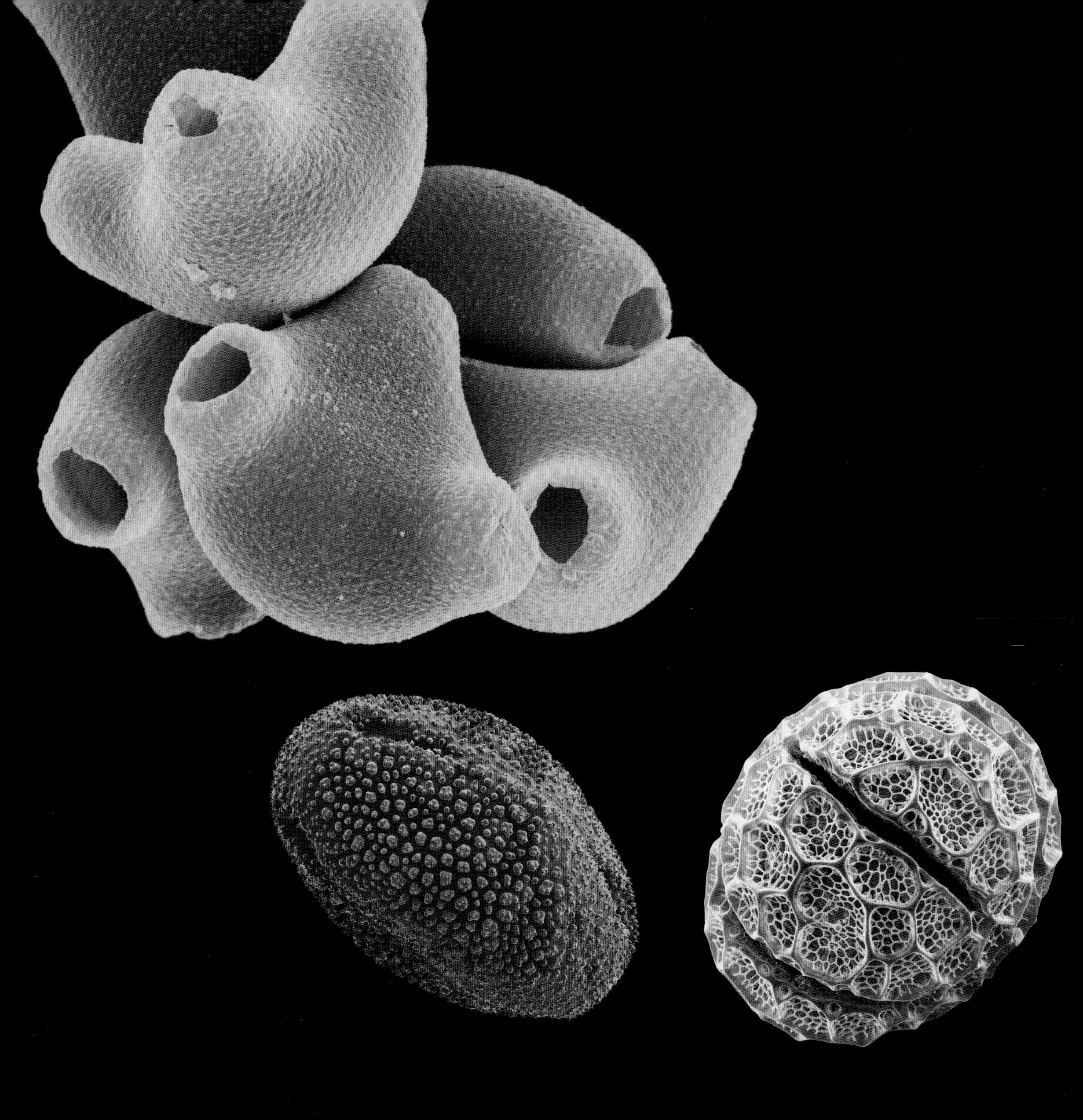

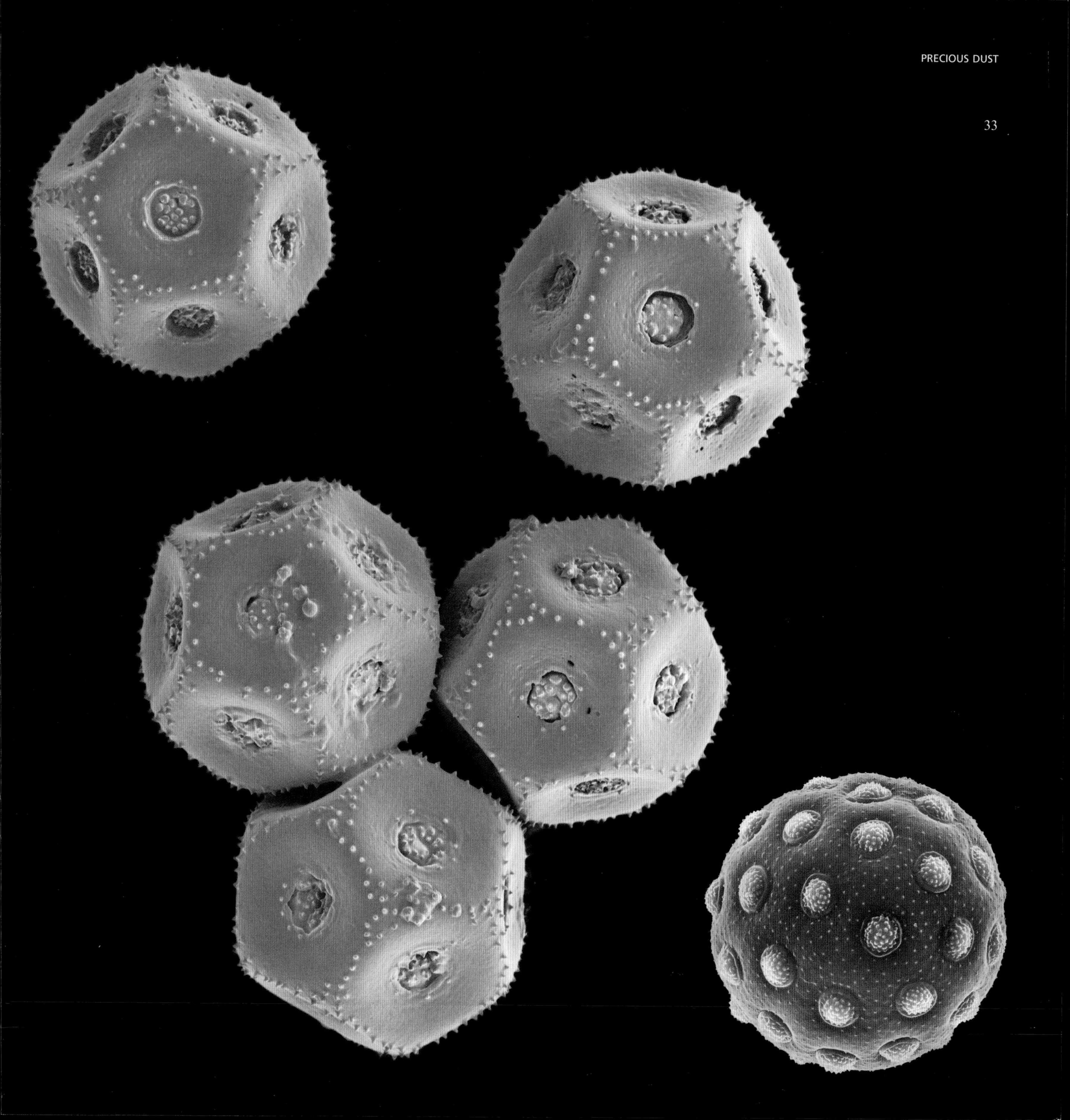

FINDING THE OTHER HALF

As marvellous as pollen is, it has one major handicap. It is unable to move independently and yet must somehow carry sperm to the receptive (female) stigmatic surface of another plant of the same species in order to avoid inbreeding. To solve this problem, plants have evolved various strategies that facilitate the transport of their pollen. These include abiotic pollination by wind or water and pollination by animals, mainly insects but also birds, and small mammals, especially bats. For a plant, simply shedding pollen into the air to be carried away by the wind to a receptive stigma of a flower of the same species is not only haphazard but also wasteful of its energy resources. It requires the production of huge amounts of pollen grains to ensure that they reach their target in sufficient numbers. Windborne pollen is often visible as clouds of yellow dust emerging from wind-pollinated plants such as Pines, Hazels, Alders, Birches and Grasses – much to the distress of hay fever sufferers. To give an idea of the numbers involved, a single maize plant (*Zea mays*, Poaceae) may produce around 18 million pollen grains.

POLLINATION BY WIND AND WATER

Wind-pollinated plants are mainly found in places where not many pollinating animals are present but plenty of wind instead. In fact, despite the high investment in producing huge amounts of pollen, wind pollination is quite cost-efficient in communities where wind-pollinated plants are common and closely spaced, for example, coniferous forests in the Arctic, the grasslands in Africa but also some broadleaf forests in the temperate regions. Our deciduous trees like Alder, Birch, Beech, Hazel, Oak, Walnut and all the Grasses are good examples of wind-pollinated angiosperms. Typical wind-pollinated plants have small flowers (large petals would be an obstacle for arriving pollen) and no scent; they are rather unspectacular (colours would be wasted on the wind) and unisexual. The male flowers are usually arranged in tassel-like inflorescences (many flowers grouped together), which shed huge amounts of very small, dry, smooth pollen grains into the air. The female flowers may be solitary or in groups but nearly always have large, feathery stigmas to catch pollen from the air.

Although only present in about two per cent of all abiotically pollinated plants, water pollination is well developed in many fresh water plants such as duckweeds (*Lemna* species, Araceae), as well as in the marine sea-grasses, flowering plants uniquely adapted to live in seawater. Sea-grasses belong to four closely related families of aquatic plants (Cymodoceaceae, Hydrocharitaceae, Posidoniaceae, and Zosteraceae). Many have strangely filamentous pollen, highly adapted to water dispersal. The pollen 'grains' of the Australian sea nymph (*Amphibolis antarctica*, Cymodoceaceae), for example, are up to 5mm long and have a similar density to seawater so that they remain submerged or float after shedding from the anther. Sea-grass pollen is released in masses, and passively carried by the tides across sea-grass flats to curl around protruding female stigmas encountered en route.

POLLINATION BY ANIMALS

Only ten per cent of plant species are wind pollinated; all others depend on the intervention of animals, mainly insects, to transport their pollen. This is for a good reason. Insects are more reliable and targeted than wind when it comes to pollinating flowers. Seeking rewards, typically in the form of pollen or nectar, animal pollinators such as bees and butterflies move from flower to flower, thus providing a relatively precise movement of pollen. Therefore, animal-pollinated flowers need much lower numbers of pollen grains to achieve pollination success, a clear reproductive benefit compared with wind-pollinated flowers. To better attach their pollen to the bodies of pollinating visitors, animal-pollinated angiosperms often have spiny or otherwise sculptured pollen. Also their pollen grains are often covered with *pollenkitt*, a sticky coating of oily lipids. Sticky pollen is one of many adaptations flowers have evolved during millions of years of co-adaptation with animal pollinators. Flowers have also developed a sophisticated range of advertising and reward strategies to attract animal couriers, but the way they 'market' their flowers depends very much on the kind of animal they want to attract.

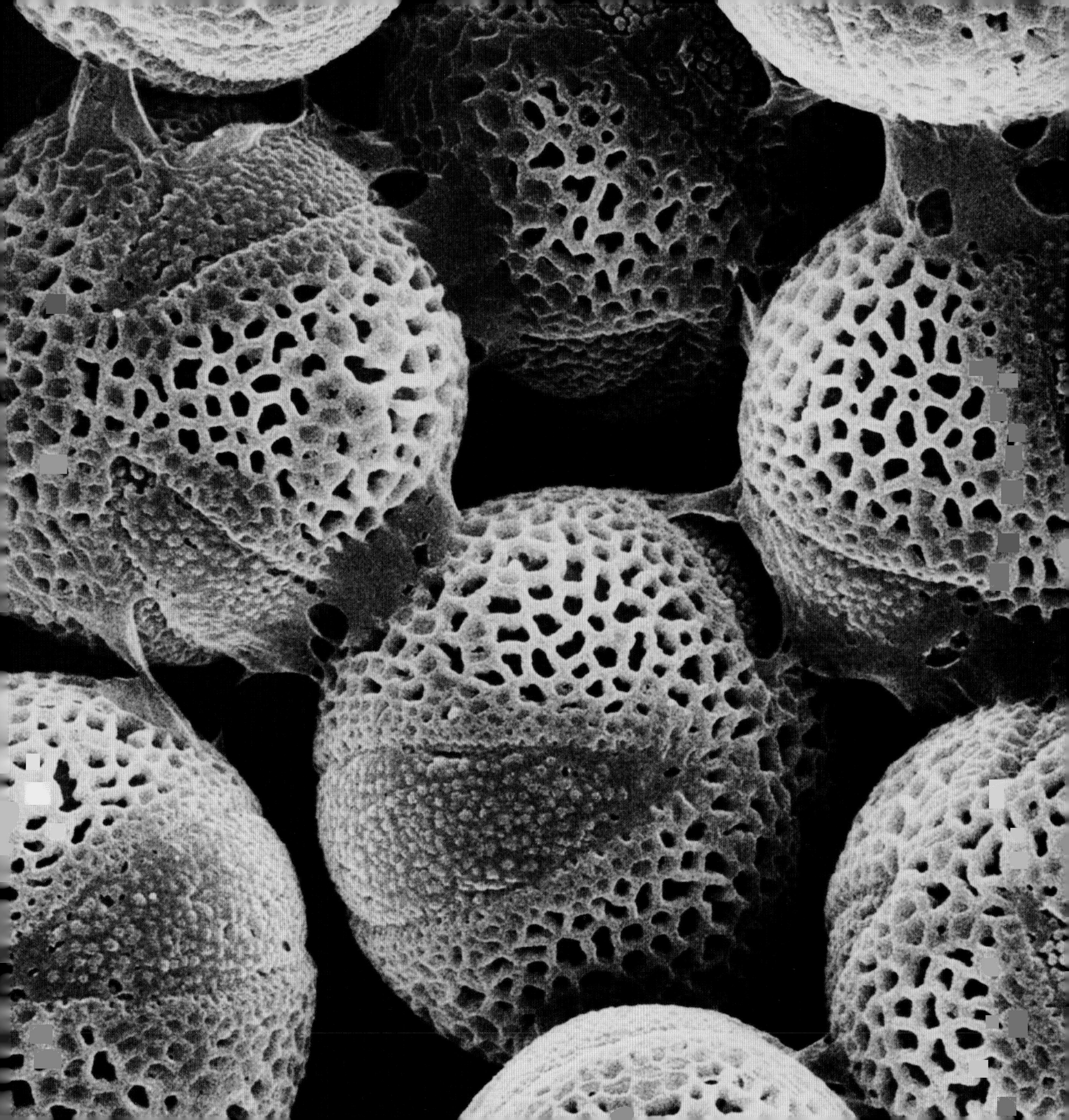

BEAUTY LIES IN THE EYE OF THE BEHOLDER

Different animals vary in body size and in their visual and olfactorial sensibilities and preferences. By co-adapting their flowers to match these preferences (for example, colours, scent, food) and physical attributes (for example, body size, length of proboscis) of particular groups, for example insects, birds or bats, or sometimes even a single species of bee, butterfly, moth or beetle, angiosperm species have developed a very efficient way of avoiding pollen of the wrong species being deposited on their stigmas. Adaptations to animal pollinators include attractants such as scent, nectar and pollen, but it is the strategic positioning of the nectar-producing organs (*nectaries*) so that the animal has to brush past the anthers and the stigma in order to reach the nectar, that is crucial. Animal attractants also include floral odours, conspicuous colour patterns (*nectar guides*) and sometimes even insect mimicry. This co-adaptation with animal pollinators is the main reason why flowering plants have developed the amazing diversity of flowers that we enjoy today. It also explains why some flowers are brightly coloured and emit a pleasant perfume (a rose or a gardenia, for example), whereas others are less pleasing, especially when they have evolved to impress carrion flies by looking and smelling like a dead animal (for example, *Amorphophallus*, *Aristolochia*, *Dracunculus*, *Rafflesia*, and *Stapelia*). Some orchids even go as far as interfering with the sex life of their pollinators by mimicking potential mating partners, for example, the Bee Orchids in the genus *Ophrys* or, perhaps even more upsetting for the poor 'suitor', a male rival that has to be attacked (for example, *Oncidium planilabre*).

Although insects are by far the most important pollinators, there are many flowers that have adapted to pollination by vertebrates, especially birds and bats, but also other small mammals and marsupials. The various adaptations that flowering plants have evolved to suit and combine with a particular type of pollinator have been described as *pollination syndromes*.

THE INSECT POLLINATION SYNDROME

Insects are not only the most ancient but also the largest group of pollinators. Over 65 per cent of angiosperms bear insect-pollinated flowers. The most important players are bees, butterflies and moths. During the course of evolution, the relationship between plants and insects developed into a very close partnership. Their alliance has become so important to both of them that not only did the plants adapt to the needs of the insect, but the insects also evolved (co-adapted) to 'fit their flowers', for example by adapting their body shape, mouth parts and foraging behaviour. In fact, the extraordinary radiation and speciation of both insects and flowering plants during the last 120-130 million years suggests that the co-adaptation between insects and flowering plants was probably the most influential factor in the origin and diversification of the angiosperms.

BEE FLOWERS

The most important group of insect pollinators are bees. There are c. 20,000 species of bees, of which the familiar European honey bee (*Apis mellifera*) is just one. Bees are very efficient pollinators and many plants have co-adapted with them to their mutual benefit. Bees are social insects: they collect nectar (as an energy source) and pollen (as a protein source for the larvae) to sustain the bee colony. The reward offered by bee flowers (i.e. flowers with a bee pollination syndrome) therefore includes both nectar and sticky, often scented pollen. Bees have a visible spectrum that excludes red but extends into the ultraviolet (UV), a colour the human eye cannot see. In order to set the flower off against the green foliage of the plant and catch the bee's attention, bee flowers are mostly yellow or blue. If they appear bright white to our eyes, they are usually strongly UV-reflective. Distinct colour patterns, called nectar guides, point the bee to the source of the nectar reward, in the same way that the white lines on a runway guide an aeroplane to a safe landing site. These nectar guides may be within the spectrum of human visibility or in the ultraviolet spectrum. Some bee flowers offer the insect a landing platform in the form of a flat, plate-like flower or inflorescence (for example, a sunflower head). Other bee flowers can have zygomorphic flowers (flowers that can be divided only by a single plane into mirror-image halves); an enlarged lower lip serves as a resting platform. Many evolutionarily advanced families such as the Figworts (Scrophulariaceae), Mints (Lamiaceae) and Plantains (Plantaginaceae) have zygomorphic flowers with fused petals that are moulded into a tube which can be accessed only by their preferred pollinators. The flowers of Snapdragons (*Antirrhinum* species, Plantaginaceae), for example, will open only for large, heavy bees or bumblebees; small bees are too light to push down the lip that obstructs the entrance to the floral tube.

BUTTERFLY AND MOTH FLOWERS

Butterflies and moths are also important insect pollinators. Both have a long tongue (proboscis), which is a specially adapted feeding/sucking food canal which curls up like a spring coil under the head of the insect when not in use. Unlike moths, which are nocturnal and have a good sense of smell, butterflies are diurnal and rely on their visual rather than their weak olfactory sense. The visible spectrum of butterflies includes ultraviolet and – unlike bees and most other insects – red. Typical butterfly flowers are slightly scented but brightly coloured; red, pink, purple and orange are butterfly favourites. As in bee-pollinated flowers, there are nectar guides. These are adapted to the way butterflies perch to feed with their long proboscises. Butterfly flowers may have a flat, plate-like landing pad and ample nectar hidden at the bottom of a long narrow tube or spur making it is inaccessible to insects with short proboscises.

Like butterflies, moths are adapted to dip for nectar in tubular flowers, their main source of food. However, since they are nocturnal animals, they are attracted by scent rather than colour. Typical moth-pollinated plants have white or pale pink flowers with no nectar guides; they open at night to emit a strong, sweet perfume often attractive to humans too. Many have long spurs adapted to certain species of moths, which helps prevent the deposit of unwanted pollen on their stigmas.

Mutual adaptation between flowers and their preferred pollinators can be so obvious that Charles Darwin predicted the pollinator of the Malagasy Comet Orchid, *Angraecum sesquipedale*, without ever having seen it. When he observed the huge 30-35cm (12-14in.) long, hollow spur inserted in the back of this flower, he postulated that an insect must exist with a tongue long enough to reach the nectar in the base of the spur, and that the insect was probably a moth. It was not until several decades after his death that Charles Darwin was proved right. In the early twentieth century, a giant Hawkmoth with a proboscis 22cm (9in.) long was discovered in Madagascar and given the Latin name *Xanthopan morganii praedicta* ('*praedicta*' meaning 'predicted'). Although it was named and described in 1903, the final proof that this Hawkmoth is indeed the pollinator of the Comet Orchid was found only 130 years after Charles Darwin's prediction. In 1992, the German zoologist Lutz Wasserthal went on an expedition to Madagascar to tracked down the elusive Hawkmoth in its natural habitat. The trip was successful and Wasserthal returned with sensational photographs which for the first time provided unquestionable evidence that *Xanthopan morganii praedicta* is indeed the pollinator of *Angraecum sesquipedale*. This still leaves the question of why this Hawkmoth developed such preposterously long mouth parts. The answer lies in the feeding strategy of Hawkmoths. Most Hawkmoths feed while hovering in front of flowers and Wasserthal believes that the extreme elongation of the insect's proboscis, and the hovering flight, are adaptations to ambushing predators, such as hunting spiders that hide among the flowers. Only Hawkmoths with extremely long proboscises can stay outside the range of hunting spiders while feeding. The likely evolutionary scenario is that Hawkmoths developed their elongated mouth parts as a defence mechanism and flowers subsequently adapted their shape to recruit suitably pre-adapted Hawkmoths as pollinators.

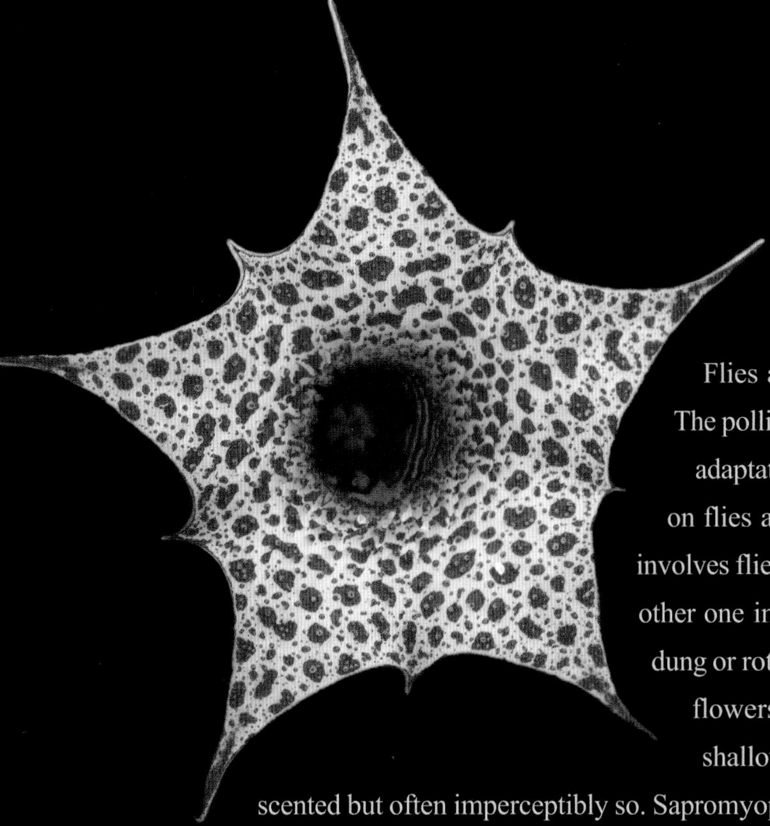

FLIES AND BEETLES AS POLLINATORS

Flies and beetles play a lesser but still important role as pollinators. The pollination syndromes evolved by certain plants as a result of their co-adaptation to these animals can be impressive, especially when they rely on flies as pollinators. There are two types of fly pollination. *Myophily* involves flies that regularly feed on pollen and nectar, such as hoverflies. The other one involves carrion dung flies and carrion flies that usually live off dung or rotting flesh, and this is also where they lay their eggs. Myophilous flowers like those of many spurges (*Euphorbia* species) are usually shallow, pale in colour and offer easily accessible nectar. They are also scented but often imperceptibly so. Sapromyophilous flowers mimic the unsavoury diet and breeding habits of dung and carrion flies by producing flowers that have the appearance and rotten smell of decaying organic matter. They are typically dull brown, medium to dark purple (e.g. *Abroma augusta* Malvaceae) or green in colour (e.g. *Deherainia smaragdina*, Theophrastaceae). Most characteristically, they emit an unpleasant, putrid smell. Certain dung- and carrion-eating beetles are also attracted to these flowers but other flowers specifically cater for pollen-eating beetles. As beetles are rather heavy, destructive visitors, such flowers are typically large, robust and bowl-shaped (e.g. *Magnolia* species, *Papaver* species, *Tulipa* species). Small flowers attract beetles if they are clustered together in dense inflorescences, as in many members of the carrot family (Apiaceae). Beetle-pollinated flowers may be odourless or emit a strong, fruity scent (e.g. the strawberry bush, *Calycanthus floridus*) and offer ample pollen as a reward but there is usually little or no nectar. Flower colour typically ranges from dull white to dark purple, but may be bright red with nectar guides as in Poppies (e.g. *Papaver rhoea)* and Tulips (e.g. *Tulipa aememsis*), which are pollinated primarily by scarabaeid beetles and, secondarily, by bees.

THE BIRD POLLINATION SYNDROME

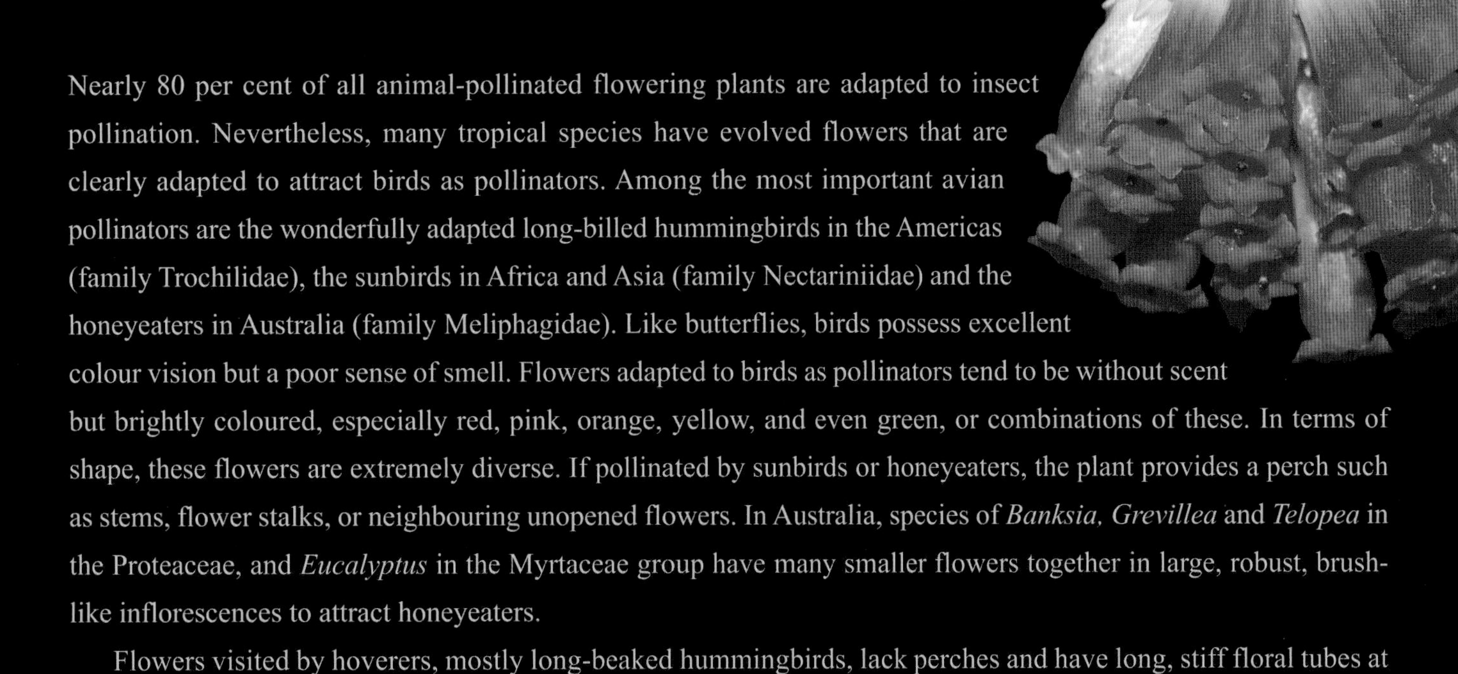

Nearly 80 per cent of all animal-pollinated flowering plants are adapted to insect pollination. Nevertheless, many tropical species have evolved flowers that are clearly adapted to attract birds as pollinators. Among the most important avian pollinators are the wonderfully adapted long-billed hummingbirds in the Americas (family Trochilidae), the sunbirds in Africa and Asia (family Nectariniidae) and the honeyeaters in Australia (family Meliphagidae). Like butterflies, birds possess excellent colour vision but a poor sense of smell. Flowers adapted to birds as pollinators tend to be without scent but brightly coloured, especially red, pink, orange, yellow, and even green, or combinations of these. In terms of shape, these flowers are extremely diverse. If pollinated by sunbirds or honeyeaters, the plant provides a perch such as stems, flower stalks, or neighbouring unopened flowers. In Australia, species of *Banksia, Grevillea* and *Telopea* in the Proteaceae, and *Eucalyptus* in the Myrtaceae group have many smaller flowers together in large, robust, brush-like inflorescences to attract honeyeaters.

Flowers visited by hoverers, mostly long-beaked hummingbirds, lack perches and have long, stiff floral tubes at the bottom of which they secrete copious quantities of easily digestible, glucose-rich nectar. Typically, hummingbird-pollinated flowers nod or dangle so that the bird has to hover below the flower and point its beak upward to insert it in the long nectar spurs. In the process, the bird's head is dusted with pollen. As in butterfly and moth pollinated flowers, there has been extensive co-adaptation between the length of the feeding apparatus and of the floral tube.

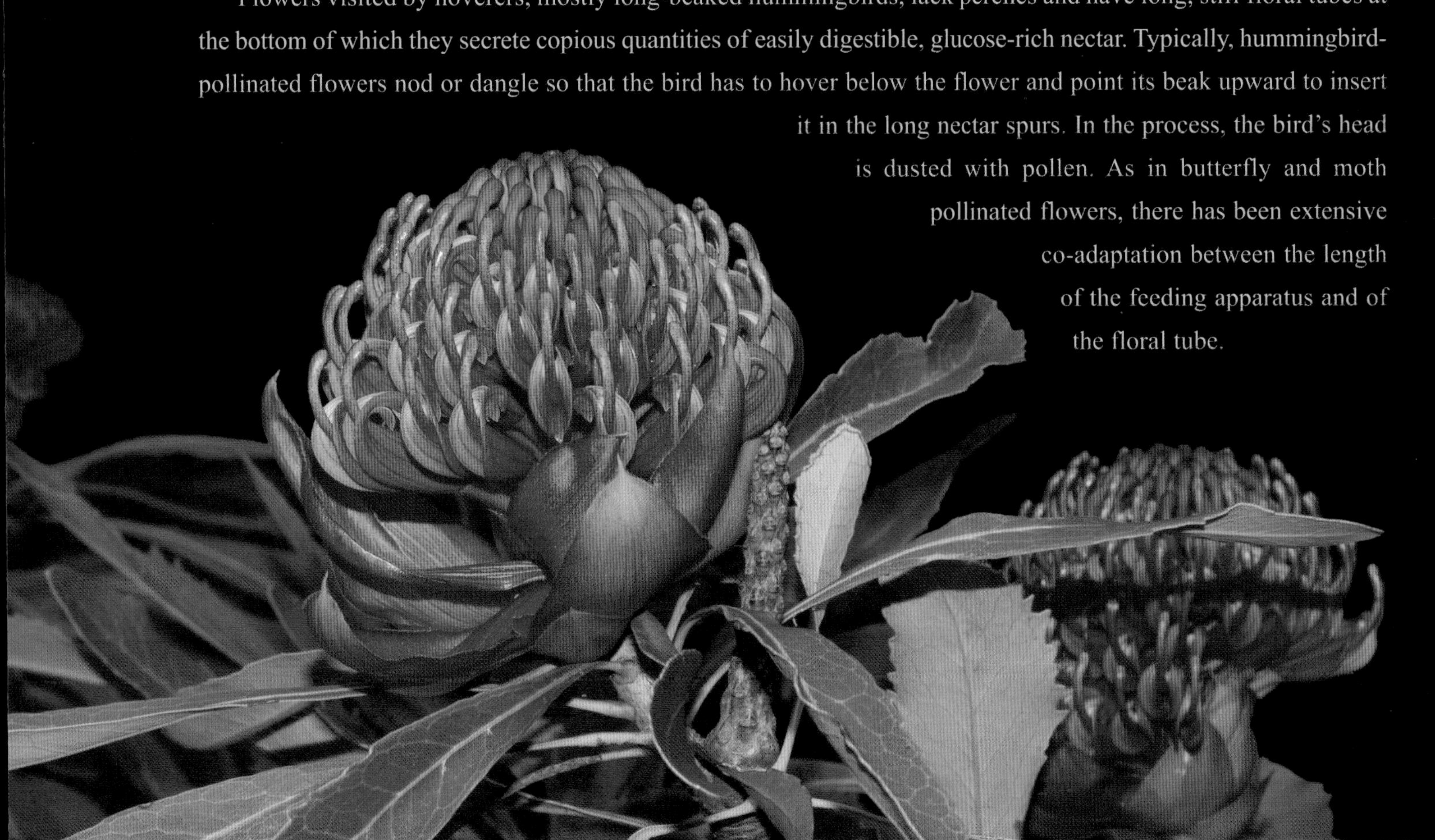

THE BAT POLLINATION SYNDROME

Among mammals, tropical bats are by far the most important pollinators. The majority of the c. 1000 species of bats that occur worldwide are insect eaters. However, among Old and New World bats, two groups have independently evolved a preference for pollen, nectar and fruit. In the tropics of the Old World it is the Fruit Bats or Flying Foxes of Pteropodidae that thrive on a vegetarian diet. They are the only family in the superorder Macrochiroptera (large bats), so named because it boasts the largest bats in the world. Although the smallest members of this family measure only 6-7cm from head to tail, Flying Foxes (*Pteropus* species) can be as long as 40cm with a wingspan of 1.7m. The Pteropodidae are widely distributed throughout the tropical and subtropical regions of Africa, Asia and Australia, and include more than 160 species. Their flower- and fruit-loving counterparts in the New World are generally smaller and belong to the family Phyllostomatidae (American Leaf-nosed Bats) in the order Microchiroptera (small bats). Contrary to Old World Fruit Bats, which have relatively simple hearing apparatus, New World Fruit Bats use sophisticated echo-location for their navigation. With just one exception (the Egyptian Fruit Bat, *Rousettous egyptiacus*), the Pteropodidae lack echo-location apparatus and rely on vision to avoid obstacles, and on their sense of smell to locate flowers and fruits. The two groups also differ slightly in their dietary preferences. Whereas Old World Fruit Bats either live entirely on pollen and nectar or fruits, their New World cousins are less strongly co-adapted to a vegetarian diet and get a large amount of their protein from insects. Typical bat flowers display a distinctive suite of characteristics. They are nocturnal, rather large in size, wide-mouthed and bell- or disc-shaped to accommodate a bat's head, robust in texture, dull in colour (white to cream or even green but sometimes pink, purple or brownish), and they have a very strong scent of cabbage or fermenting fruit and produce large quantities of watery nectar. Bat flowers are typically borne outside the foliage to be easily accessible, either directly on the trunk and larger branches or on long, stalks that hang down from the branches.

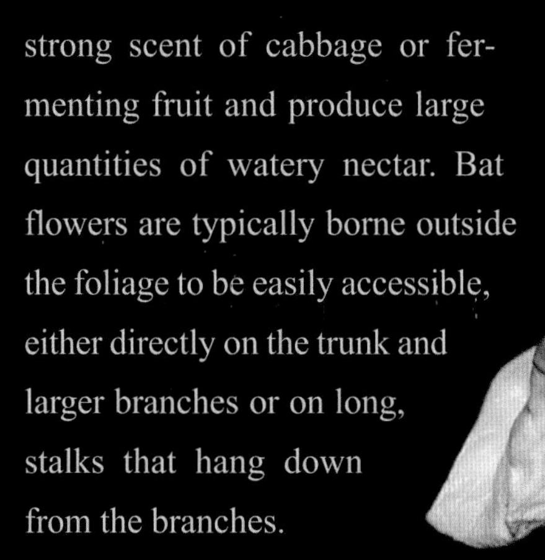

TYPICAL BAT FLOWERS

Good examples of bat-pollinated species are found in the Bignoniaceae, for example, the African Sausage Tree (*Kigelia pinnata*) and the American Calabash tree (*Crescentia cujete*); other examples include the Cup-and-Saucer Vine (*Cobaea scandens*) in Polemoniaceae from tropical America, and many columnar cacti (family Cactaceae) such as the Saguaro (*Carnegiea gigantea*), Cardón (*Pachycereus pringlei*) and the Organ Pipe Cactus (*Stenocereus thurberi*). Bat flowers can also be of the brush or pincushion type: large flowers, or inflorescences of clustered flowers with a large number of showy stamens that often serve as food for the bats (instead of nectar). The flowers of the African Baobab (*Adansonia digitata*, Malvaceae), for example, have up to two thousand stamens. As with other pollination syndromes, not every bat-visited flower corresponds perfectly with the bat-pollination syndrome and the flowers of many species are visited by a range of different pollinators. For example, *Kigelia africana* has blood-red, nocturnal flowers that are attractive not only to bats but also to moths and sunbirds.

In the species-rich habitats of the tropics and subtropics, various other small mammals may transfer pollen as they forage. The Malagasy Traveller's Palm (*Ravenala madagascariensis*, Strelitziaceae), though mainly bird-pollinated, is also adapted to Lemur pollination. In Hawaii, small nocturnal rats, called white-eyes, are reported to climb around *Freycinetia arborea* trees (Pandanaceae) to nibble the juicy bracts of their inflorescences, which the plants have produced to attract Fruit Bats. In Australia there are a number of small marsupials that transfer pollen while foraging. Some show no signs of adaptation as pollinators, while others, such as the Honey Possum (*Tarsipes rostratus*), are highly adapted pollinators. The Honey Possum, which feeds on the nectar of the long narrow flowers of the Protea family (Proteaceae), has a strongly projecting snout, very few teeth or even none at all, and a long, narrow tongue with a brush-like tip.

One of the most bizarre pollination mechanisms occurs in the flowers of the Nippon Lily (*Rohdea japonica*, Ruscaceae) from China and Japan. They smell of rotting bread and attract slugs and snails, which feed on the fleshy flowers and, as they creep around, transfer the pollen stuck to their slimy bodies. Pollination by snails and slugs is rare and has been discovered in only six other plant species, mainly members of the arum family (Araceae; for example, *Calla palustris*, *Colocasia odora*, *Philodendron pinnatifidum*, *Lemna minor*).

THE ADVANTAGES OF ANIMAL POLLINATION

With their own private courier service, flowering plants are able to avoid hybridization (inbreeding) with close relatives. This very efficient genetic isolation mechanism allows many new species to evolve within a relatively short time, even in close proximity to their first and second cousins. A pollinator adapted to a particular flower may travel long distances between two flowers of the same species. This allows plant communities to be more diverse, by having a higher species number but a lower number of individuals of each species in a given space. The best example of this strategy in action is well demonstrated by orchids. They possess the most sophisticated flowers of all angiosperms and with more than 18,500 species, are the largest and most successful group of flowering plants on Earth. Only their extremely selective pollination mechanism could make it possible for more than 750 species of orchids to co-exist on one mountain, Mount Kinabalu (Borneo). However pollination is achieved, once its ovules are fertilised, a flower prepares to become a fruit: its petals wilt and drop, the ovules enlarge and begin to turn into seeds, the ovary starts to grow to let the developing seeds expand, and the ovary wall will become the fruit wall (*pericarp*).

FRUITS AND SEEDS

The word *fruit* brings many things to mind: crunchy apples, sweet cherries, aromatic strawberries and tropical delicacies like bananas, pineapples and mangoes. There are some 2,500 edible tropical fruits worldwide but most are used only locally by indigenous people. But, whether they are tropical, subtropical or temperate, we enjoy fruits in many ways, either fresh, dried, cooked or preserved, in yoghurts, ice creams, jams and biscuits; as juices, coffee or alcoholic drinks. Some are used as spices: peppercorns, nutmeg, cardamom, cloves and chilli peppers. The most valuable of all are the fermented pods ('beans') of the Vanilla Orchid (*Vanilla planifolia*) a highly priced flavouring in chocolate, ice cream and many other sweet dishes. Others, such as the fruits of the West African Oil Palm (*Elaeis guineensis*) and Olive (*Olea europaea*) are pressed for their valuable oils. Countless other fruits are important to humans as a source of natural raw materials such as fibres, dyes and medicines, or simply as decoration.

For us, fruits are indeed a wonderful gift from nature and provide us with delicacies and an extraordinary range of useful commodities. But none of the above tells us why plants produce so many different fruits. Besides, many fruits are inedible, either because they

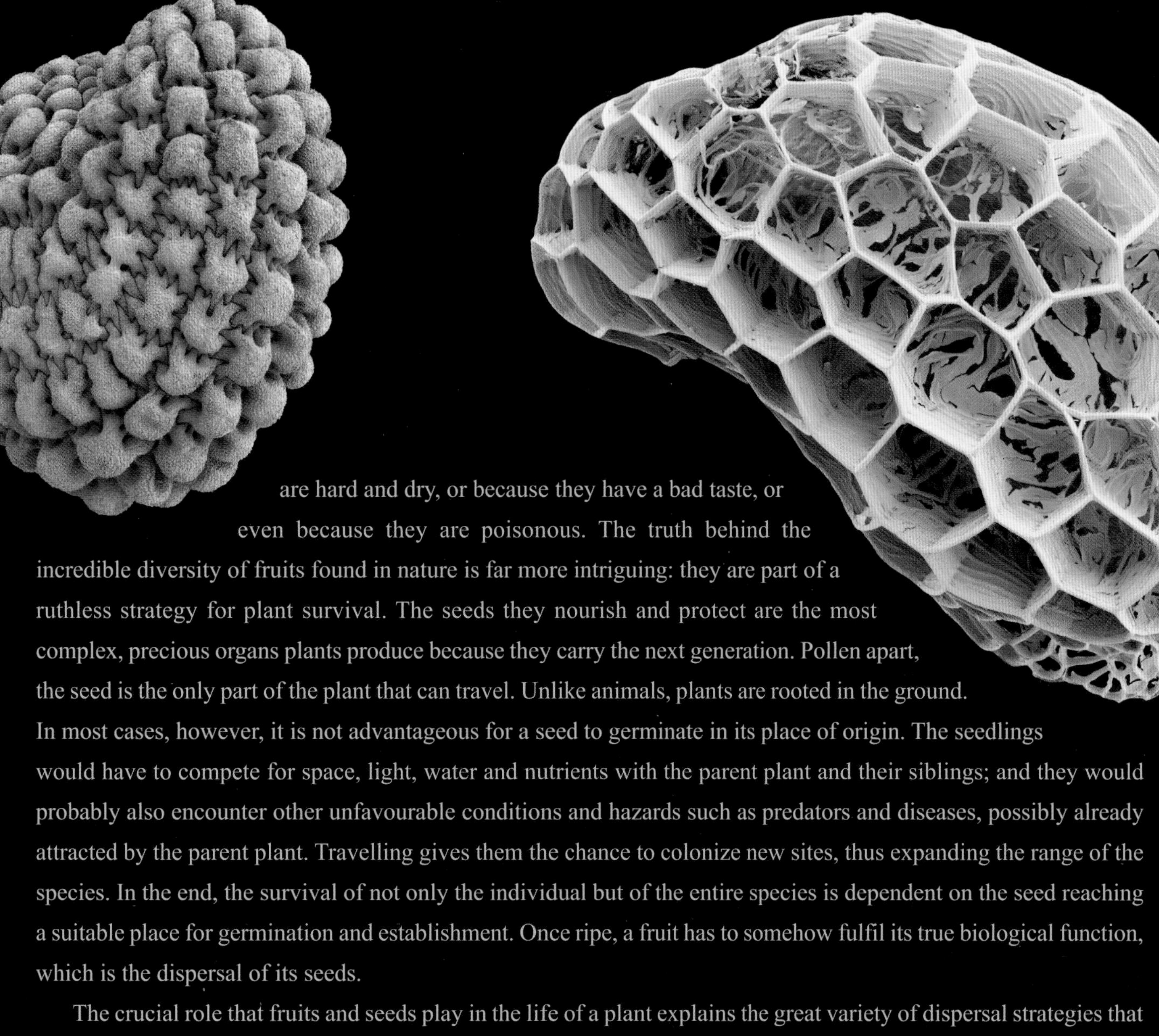

are hard and dry, or because they have a bad taste, or even because they are poisonous. The truth behind the incredible diversity of fruits found in nature is far more intriguing: they are part of a ruthless strategy for plant survival. The seeds they nourish and protect are the most complex, precious organs plants produce because they carry the next generation. Pollen apart, the seed is the only part of the plant that can travel. Unlike animals, plants are rooted in the ground. In most cases, however, it is not advantageous for a seed to germinate in its place of origin. The seedlings would have to compete for space, light, water and nutrients with the parent plant and their siblings; and they would probably also encounter other unfavourable conditions and hazards such as predators and diseases, possibly already attracted by the parent plant. Travelling gives them the chance to colonize new sites, thus expanding the range of the species. In the end, the survival of not only the individual but of the entire species is dependent on the seed reaching a suitable place for germination and establishment. Once ripe, a fruit has to somehow fulfil its true biological function, which is the dispersal of its seeds.

The crucial role that fruits and seeds play in the life of a plant explains the great variety of dispersal strategies that plants have developed during the course of evolution. These functional adaptations can be obvious and aesthetic – for example, the wind-adapted winged fruits of Sycamore and Ash Trees – or structures that resemble sophisticated pieces of engineering. It is not surprising that the dispersal of fruits and seeds has long fascinated biologists and non-biologists alike. Their strategies for dispersal – whether involving wind, water, animals and humans or an explosive force from the plant itself are reflected in a seemingly endless plethora of different colours, sizes and shapes.

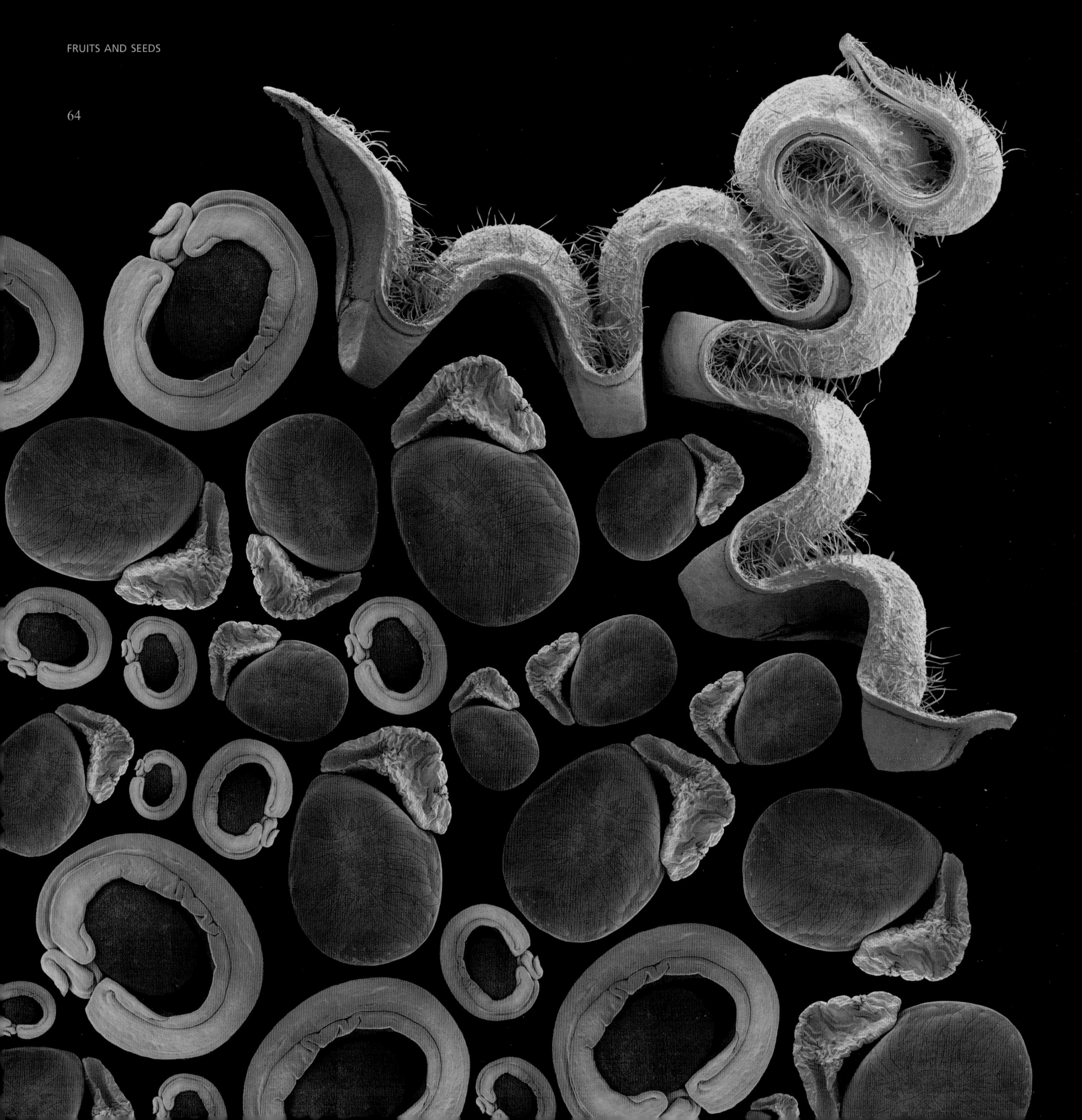

THE VARIOUS WAYS TO GET AROUND

Fruits either open to release their seeds into the environment (dehiscent fruits) or remain closed even when ripe (indehiscent fruits). Depending on the type of fruit, the nature of the dispersal unit, the *diaspore*, varies. In *capsules* and other dehiscent fruits, it is the seed itself that functions as the diaspore. In indehiscent fruits such as *berries* (with their fleshy fruit wall), *nuts* (with their hard, dry fruit wall) or *drupes* (with a fleshy outer fruit wall , and a hard stone around the central seed), the diaspore constitutes the entire fruit. There are also fruits in which the ripe fruit represents a whole mature inflorescence or *infructescence*. Familiar examples of such *compound fruits* include the pineapple (*Ananas comosus*, Bromeliaceae) and some delicious members of the mulberry family (Moraceae) including the mulberry itself (*Morus nigra*), figs (*Ficus carica*) and – most remarkably – the tropical jackfruit (*Artocarpus heterophyllus*). At a length of up to 90cm and a weight of up to 40 kilos, the jackfruit qualifies as the largest fruit borne by any tree on Earth.

Diaspores may also consist of seeds or entire fruits or infructescences or fruit fragments: in Maples (*Acer* species, Sapindaceae) the fruits remain closed but break into two pieces, called *fruitlets*. Whatever the nature of their diaspores, plants pursue four principal strategies of dispersal: they can either rely on natural processes (wind and/or water dispersal), have fruits that actively self-disperse their seeds, or, by adaptation, persuade or lure animal couriers to work for them (animal dispersal). The great variety of diaspores found among seed plants is predominantly the result of adaptations to these four dispersal mechanisms. The dispersal strategy of a diaspore is usually reflected in its appearance and revealed by its shape, colour, consistency and size.

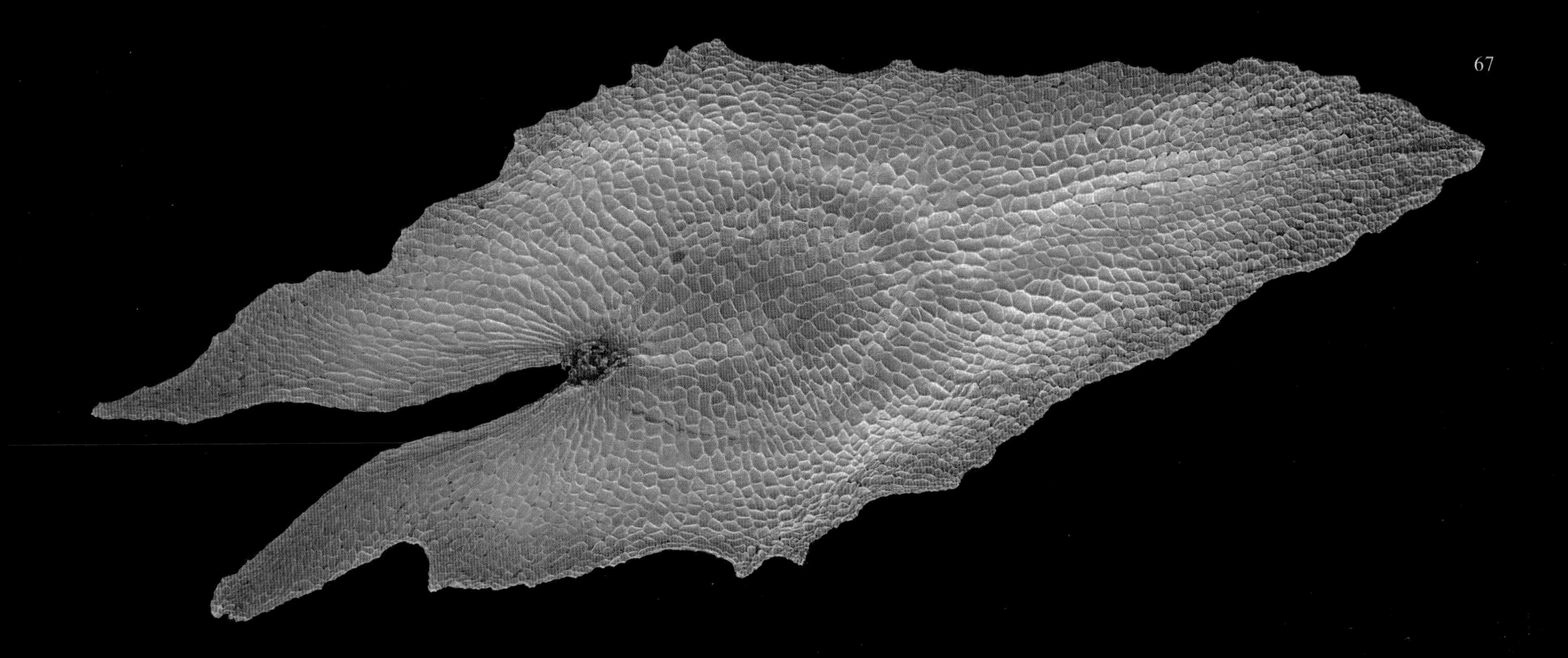

WIND DISPERSAL

Some of the most obvious adaptations of diaspores are those that facilitate wind dispersal (*anemochory*). Wings, hairs, feathers, parachutes or balloon-like air chambers are tell-tale signs of a wind-dispersal syndrome. These structural specialisations increase the aerodynamic properties or air buoyancy of the diaspores. They can be present in the seeds, or in the fruits themselves if they are indehiscent. Whichever organs are involved, the tissues from which these structures are formed usually consist of dead, air-filled cells with thin walls to ensure minimum weight.

Despite the fact that wind would seem an unreliable and unpredictable agent to be put in charge of one's offspring, anemochory has certain advantages. Air currents can be very strong and a storm can carry a fruit or a seed far away, sometimes over many kilometres. Travelling on the wind is also economical since the energy-rich rewards needed to attract animal dispersers are unnecessary. The great disadvantage of wind dispersal is that the distribution of the diaspores is dependent on the direction and strength of the wind. Wind dispersal is therefore haphazard and potentially wasteful. Most wind-dispersed seeds are doomed because they fail to reach a suitable place where they can grow into a new plant. At least some of the energy that is saved by not having to produce rewards for more reliable animal dispersers has to be invested in the production of a larger number of seeds to allow for wastage.

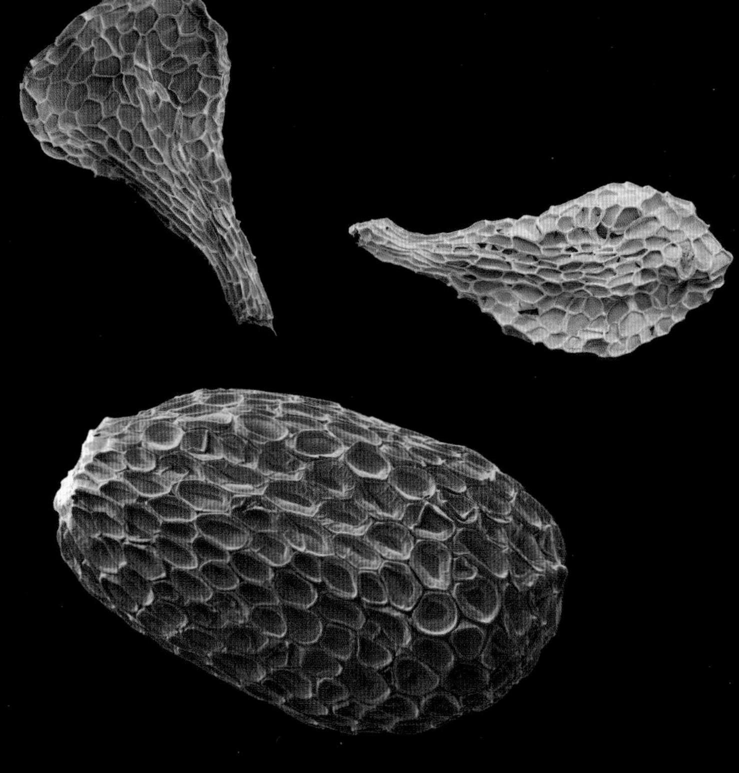

SEEDS LIKE DUST

The most effective strategy to ensure long-distance dispersal by the wind is the production of a large number of extremely small and lightweight seeds. To give an impression of the dimensions involved, a single capsule of the tropical American orchid *Cycnoches chlorchilon* produces almost four million seeds; and one gram of the smallest wind-dispersed orchid seeds (for example, *Calanthe vestita*) contains more than two million seeds. The large surface-volume ratio of such 'dust seeds' significantly reduces their sink velocity in air; for example, a small orchid seed sinks at about 4cm/sec, which is slow compared to the samara of an elm, which falls at a speed of 67cm/sec. Air buoyancy is further increased by a syndrome of obvious adaptations such as air pockets. Seed air bladders may consist of large, empty cells, intercellular spaces, or a space between the seed coat and the embryo-bearing centre of the seed. Seeds equipped with such air spaces are commonly called 'balloon seeds'. Typical dust seeds, without air chambers, are those of the Broomrapes – *Orobanche* species (Orobanchaceae), Sundews (Droseraceae) and many members of the heath family (Ericaceae) such as *Erica* species and *Rhododendron* species. Balloon seeds are most famously found in orchids but also in many other plants such as Butterworts (*Pinguicula* species, Lentibulariaceae), Foxgloves (*Digitalis* species, Plantaginaceae), and certain Loasaceae (for example, *Loasa chilensis*).

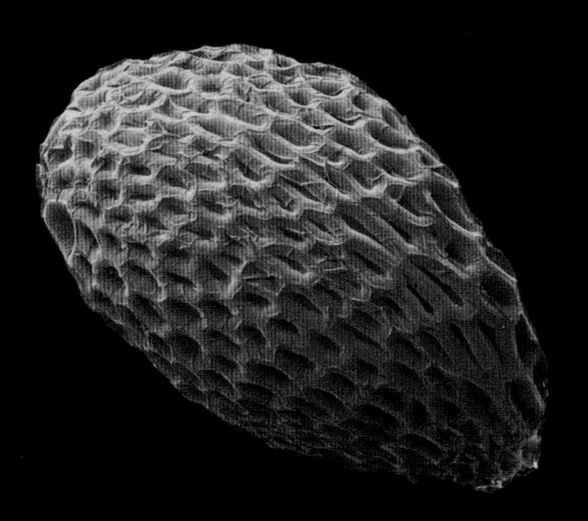

MASTERPIECES OF NATURE

The seed structures that plants have evolved as adaptations to wind dispersal are aesthetically pleasing and often masterpieces of engineering. The tiny dust and balloon seeds described above show some of the most spectacular of all seed structures, but their incredible sophistication is only revealed at high magnification. Although largely unrelated, families with this type of seed may display striking convergences. In many, if not most of the families involved, the single-layered seed coat has a distinct honeycomb structure with either isodiametric or elongated facets. The honeycomb pattern ensures maximum stability with minimum thickness and weight for load-carrying parts. Honeycomb patterns can be observed in both the inanimate and animate world: for example, in the arrangement of the carbon atoms in graphite, or the honeycombs of bees. They are also present in the surface structure of pollen in some plant species, and are applied in modern construction engineering where honeycomb cores afford stability to sandwich structures (e.g. doors and other lightweight components for aircraft). In the case of seeds, the honeycomb pattern is created by the dead, air-filled cells of their single-layered seed coat. The radial walls are slightly thickened whereas the outer and, in extreme cases, the inner tangential walls remain thin and collapse as the cells dry out. This not only reveals the intricate honeycomb pattern of the seed coat but also greatly increases the surface area of the seed and its air resistance and buoyancy.

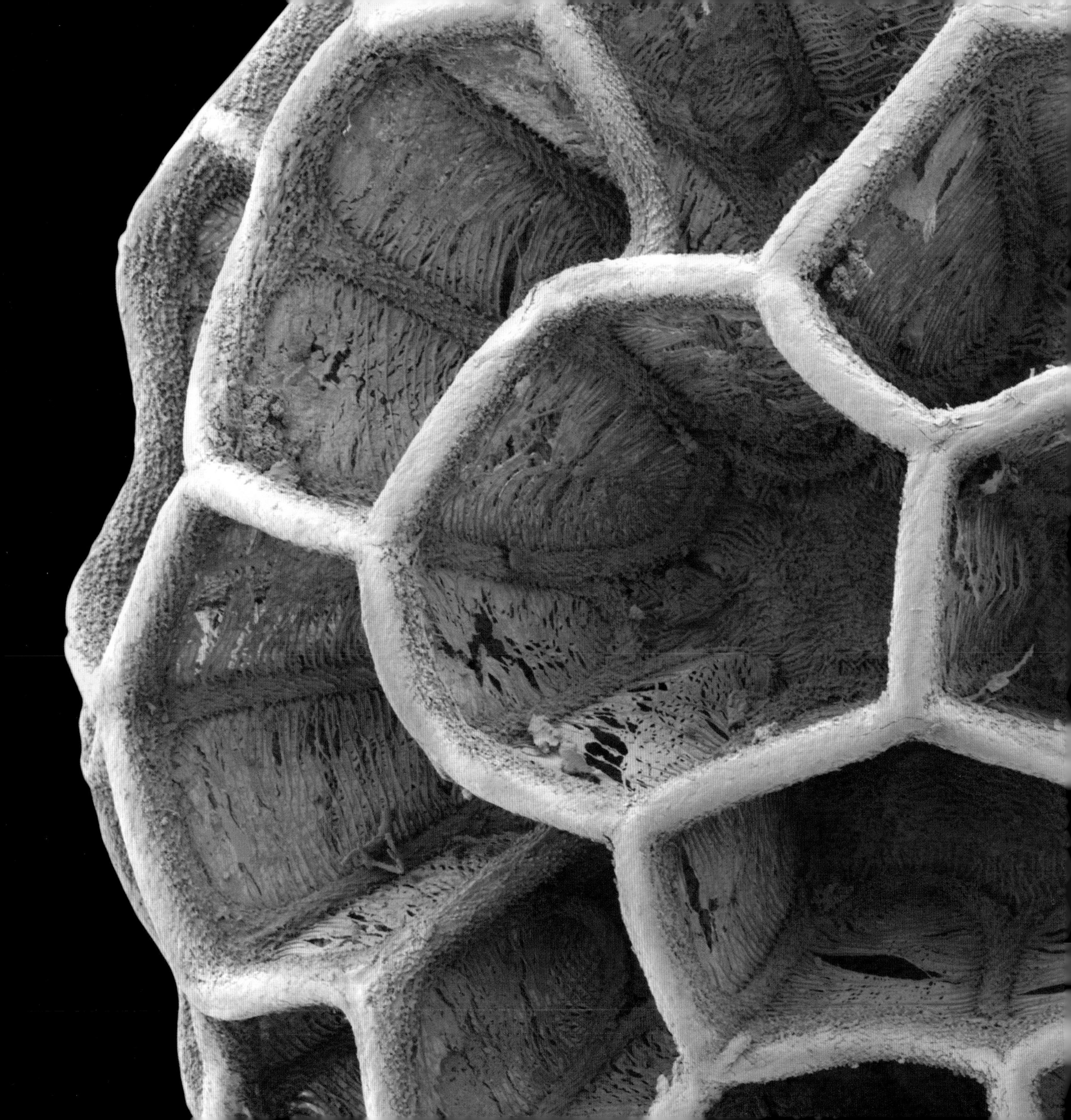

INDIRECT WIND DISPERSAL

Wind can effect the dispersal of seeds indirectly by blowing dehiscent fruits so that they scatter or eject their seeds. This type of wind dispersal is called *anemoballism* and involves many herbaceous plants with capsules on long, flexible stalks. In Poppies (*Papaver* species, Papaveraceae) the ring of pores around the top of the capsules functions like a pepper pot as they sway in the wind, releasing large numbers of tiny seeds. The circle of narrow openings through which the seeds leave the capsule is well protected from the rain by the protruding rim of the capsule 'lid', the platform that carries the remains of the papillose stigma. Pinks such a *Petrorhagia nanteuilii*, Campions (*Silene* species), Carnations (*Dianthus* species) and Primroses (*Primula* species) follow the same strategy but their open capsules have tiny teeth at the apex, leaving only a narrow opening for the seeds to escape. In the curious capsules of Snapdragons (*Antirrhinum* species, Plantaginaceae) three irregular apical pores with tiny recurving valves burst open. In the Lesser Snapdragon (*Antirrhinum orontium*) the long style persists and develops into a projecting rod, perhaps so that passing animals shake them even more effectively than the wind. Although the seeds of many anemoballistic plants have highly ornate surface patterns, this is not usually accompanied by any distinct anatomical modifications that could further assist dispersal. However, their small seed size allows them to travel considerable distances if they are ingested by browsing livestock or stick to their hoofs.

WATER DISPERSAL

Water facilitates dispersal in a variety of ways. The air bladders of balloon fruits and seeds and the high surface/weight ratio of many small wind-dispersed diaspores coincidentally also afford good buoyancy in water. Plumed and winged diaspores, if small enough, can also stay afloat thanks to the surface tension of water; the tiny winged seeds of *Spergularia media*, for example, are able to float for many days. Nevertheless, water dispersal of otherwise anemochorous diaspores is a chance event. Specific adaptations to water dispersal (*hydrochory*) are found in aquatic plants, marsh and bog plants, and other plants growing close to water. The most important property of water-dispersed diaspores is, of course, buoyancy, which is often enhanced by a water-repellent surface. Impermeability to water also inhibits premature germination of the seed and provides protection against salt water in sea-dispersed diaspores. Buoyancy is most commonly increased by enclosed air spaces and waterproof corky tissues.

RAFTERS AND SAILORS

Water-dispersed diaspores often have hooks or spines which help them to anchor on a suitable substrate or to attach themselves to the fur or feathers of animals. The seeds of the aquatic *Nymphoides peltata* (Yellow Floatingheart, Menyanthaceae) combine several of these adaptations. Once the fleshy parts of the fruit have rotted or been eaten by snails they open at the base to release the seeds directly into the water. The flat, discoid shape, the fringe of bristles around the periphery and their water-repellent surface enable the seeds to use the surface tension of water to avoid sinking. Although they are heavier than water, the seeds can stay afloat for two months if they are not disturbed. Their bristles also mean that several seeds can become entangled and form small chains or rafts on the surface of the water where they can also easily hook on to water birds to 'hitch a ride'.

Tropical islands and coastal areas are rich in plants with fruits that are able to travel in the salty water of the sea. The seeds and fruits of many plants growing on or near the coast eventually end up in the sea and get carried away by ocean currents. Fruits and seeds may be shed directly on the beach or drop into tidal pools and swamps, from where they are swept away by the tide. If they originate further inland they can reach the sea via streams and rivers. In many cases this happens accidentally.

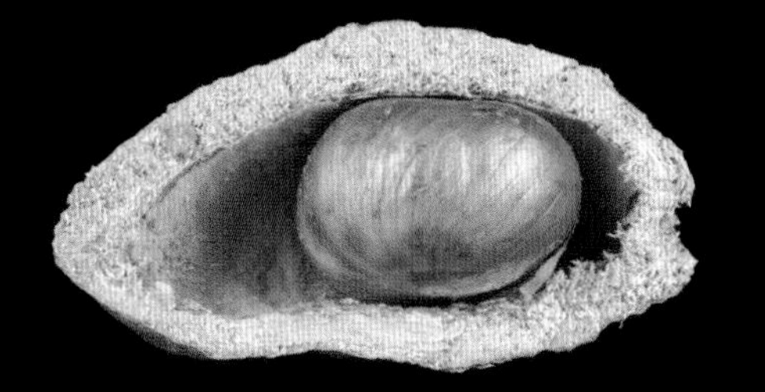

However, a number of plants, especially in the tropics, have diaspores that are specifically adapted to travel in seawater for months or even years. The waterproof nut-like fruits of the Looking Glass Mangrove (*Heritiera littoralis*, Malvaceae) are up to 10cm long and contain a single round seed surrounded by a large air space that helps keep them afloat. Astonishingly, along their back the fruits possess a prominent keel, which acts like the sail of a sailing boat while they are at sea. Other tropical fruits adapted to sea dispersal are drupes, which possess a thick, cork-like floating tissue. Fruits of this type are found in palms such as the Nipa Palm (*Nypa fruticans*) and the coconut (*Cocos nucifera*). The Nipa Palm is very common in mangrove swamps and tidal estuaries around the Indian Ocean and the Pacific. When ripe, the huge, football-size fruits break up into obovoid angular fruitlets. The single seed inside each fruitlet germinates before it is dispersed, and the emerging pointed shoot assists in the detachment. With their hard outer epicarp and underlying fibrous-spongy mesocarp the fruitlets of the Nipa Palms are well adapted to seawater. However, no fruit proves the success of this model better than coconuts the most accomplished 'sailors' among all seaworthy fruits. Coconuts are perfectly adapted to seawater dispersal, can endure months of travelling on the ocean currents, and can cover distances of up to 5,000 kilometres. This ability to travel such long distances explains the presence of Coconut Palms all over the tropics.

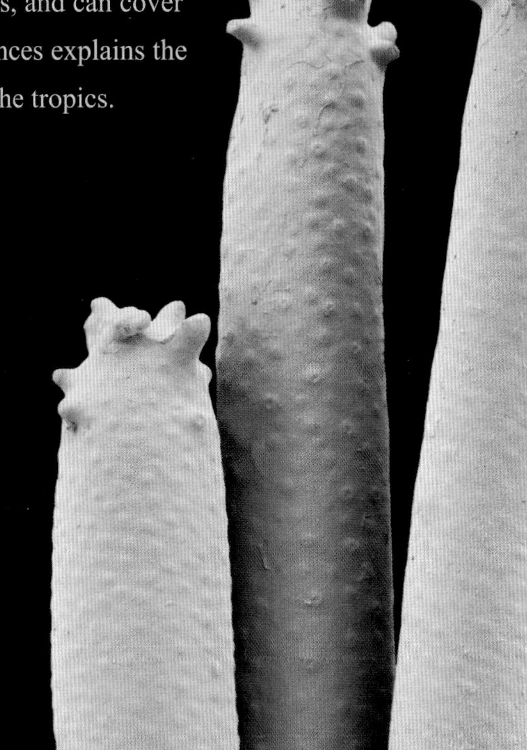

SEA BEANS

There are a number of other seagoing diaspores that can reach the main ocean surface currents and be carried thousands of miles away from their place of origin. Charles Darwin was thrilled by the thought that seeds from tropical countries could travel to European shores. Fruits and seeds from South America and the Caribbean are regularly carried by the Gulf Stream to the beaches of Northern Europe where they find themselves in a rather inhospitable environment. The most frequent arrivals from the New World are seeds belonging to members of the bean family (Leguminosae), which probably explains why they are called 'sea beans'. They clearly did not come from any local plant and, throughout history, have appeared very exotic to people, especially in the Middle Ages, when the mystery surrounding their origin inspired many legends and superstitions. The people of Porto Santo, an island in the Azores, still call the exotic 'Sea Heart' drift seed of *Entada gigas* (Fabaceae), 'Fava de Colom' (Columbus's Bean) because they believe that Christopher Columbus was inspired by finding one of these exotic beans washed up on a Spanish beach. *Entada gigas* is an enormous liana that grows in the tropical forests of Central and South America and in Africa. Its seeds are among the most commonly found drift diaspores of European beaches. The big brown, heart-shaped seeds have a diameter of up to 5cm and are carried in the largest of all legume pods, measuring up to 1.8m in length. Sea hearts, and the large seeds of the related *Entada phaseoloides* from Africa and Australia, were carved into snuff boxes and lockets in Norway and other parts of Europe. In England, the seeds were used as teething rings and good luck charms for children to protect them at sea. Even today, sea beans are valued by collectors and creators of botanical jewellery for their beautiful shapes and colours. Apart from the Sea heart, the most famous sea beans include the true sea beans (*Mucuna sloanei* and *M. urens*), the Sea Purse (*Dioclea reflexa*), yellow and grey Nickernuts (*Caesalpinia major* and *C. bonduc*). Grey Nickernuts were worn as amulets by the people of the Hebrides to ward off the Evil Eye. It was said that if the seed turned black the wearer was in danger. Lastly, there is Mary's bean (*Merremia discoidesperma*) which belongs to the Morning Glory family (Convolvulaceae) and probably has the most intriguing history of all sea beans. Produced by a woody vine that grows in the forests of Southern Mexico and Central America, it has black or brown, globose to oblong seeds 20-30mm in diameter. Its hallmark is a cross formed by two grooves, hence the name 'Crucifixion Bean' or 'Mary's Bean'. For Christians the seeds had a special symbolic meaning. If the bean could survive an ocean voyage it was believed that it would protect anyone who owned it. In the Hebrides they were reputed to ensure an easy delivery to a woman in labour and became precious talismans handed down from mother to daughter for generations.

THE BIGGEST SEED IN THE WORLD

The most enigmatic of all drift fruits is the Seychelles nut, the fruit that carries the largest seed in the world. The Seychelles nut is not closely related to the coconut but is similar and often called a 'Double Coconut' or 'Coco de Mer'. Unlike the coconut, however, it cannot float when fresh or survive prolonged contact with seawater. Nevertheless, since the fifteenth century, long before the Seychelles were discovered in 1743, the endocarps were found washed up on the beaches of the Indian Ocean. Since most of them were found in the Maldives the species was given the somewhat misleading Latin name *Lodoicea maldivica*. The true distribution of this extraordinary palm tree is limited to two of the Seychelles islands, Praslin and Curieuse. What made the Seychelles nut so famous was not only its size but also the rather suggestive shape of the fruits. Reminiscent of female buttocks, they gave rise to some superstitious beliefs. Malay and Chinese sailors thought that the double coconut grew on a mysterious underwater tree similar to a coconut palm. In Europe the highly prized fruits were thought to possess medicinal properties and their endosperm was believed to be an antidote against poison. The Seychelles nut palm may have the largest seed in the world but it contains only a very small embryo embedded in a large endosperm (nutritious tissue). The world record for the largest embryo of any seed plant is held by a member of the legume family. The seed of *Mora Megistosperma* (syn. *Mora oleifera*), a large tree from tropical America, can be up to 18cm long, 8cm wide, and weigh up to a kilogram. The bulk of the seed consists of two thickened cotyledons just as in the seeds of more familiar legumes like beans, peas and peanuts. The only difference is that the seeds of *Mora megistosperma* have an air-filled cavity between the cotyledons that affords them buoyancy in sea-water, an adaptation to their tidal marshland habitat.

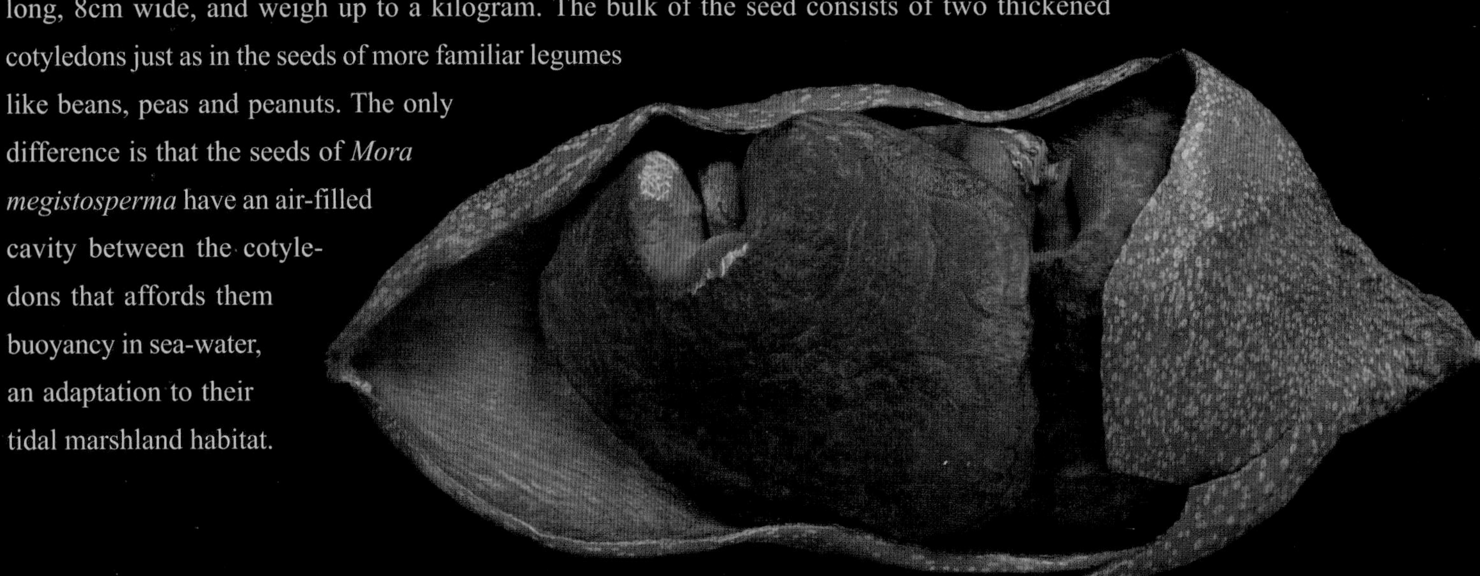

EXPLOSIVE STRATEGIES

Some plants have developed mechanisms enabling them to self-disperse their seeds. Self-dispersing seeds may not sound very advanced but self-dispersal or *autochory* involves highly complex mechanisms for plants to catapult their seeds. *Ballistic dispersal* involving explosive opening mechanisms in fruits can be triggered either by passive (hygroscopic) movements of dead tissue drying out or, in living cells, from activity caused by high hydraulic pressure. There are many familiar examples in the bean family (Leguminosae) that never fail to attract curious children, notably Lupins (*Lupinus* species), Gorse (*Ulex europaeus*) and Sweet Peas (*Lathyrus odoratus*). During the build up to 'detonation', the two halves of the fruit twist in opposite directions until they suddenly separate and eject their seeds. The seeds are only dispersed over a short distance of two metres or less. More powerful self-dispersers occur in the tropics. The Sandbox Tree (*Hura crepitans,* Euphorbiaceae) has mandarin-size fruits which, when ripe, erupt with great force ejecting their seeds up to 14m. In the Witch Hazel family (Hamamelidaceae) a specialised endocarp facilitates ballistic dispersal; the capsules open slowly but, once open, further desiccation makes the hard endocarp change its shape and grip the single seed in each of the two loculi like a vice. Rising pressure then pushes the hard, smooth seed out in a ballistic trajectory. Fleshy fruits that actively build up hydraulic pressure in living tissue until they detonate at the slightest tremor include the segments of the fusiform capsules of Touch-me-not (*Impatiens species*, Balsaminaceae), which curl up instantly and hurl seeds in all directions. At their breaking point the fruits are so sensitive to touch that anything from a passing animal, to the wind, or even flying seeds from a neighbouring fruit can trigger an explosion. The gherkin-size fruits of the Squirting Cucumber (Cucurbitaceae) squeeze their seeds in a good supply of lubricating watery liquid through a narrow basal opening, formed as the fruit stalk pops out like a champagne cork.

ANIMAL COURIERS

Wind and water dispersal are advantageous in certain habitats and suit the lifestyles of many plants. For example, in the temperate deciduous forests of North America about 35 per cent of all woody plants produce wind-dispersed fruits or seeds. Nevertheless, dispersal by wind or water is wasteful and unpredictable because the strength, direction and frequency of air and water currents are variable and unreliable. When seeds are scattered randomly,they frequently end up in places unsuitable for germination and so go to waste. Ballistic dispersal is equally random and the dispersal distances are short. Seed dispersal by animals – like animal pollination – eliminates much of the uncertainty associated with abiotic dispersal agents and offers far more efficient options. Animals follow certain behavioural patterns, which makes their movements less haphazard than those of wind and water and so animal dispersal or *zoochory* is much less wasteful than dispersal by wind or water. Plants that have adapted their diaspores to animal dispersal need to produce fewer seeds in order to guarantee the survival of their species. The ability to save costs in terms of energy and 'building materials' affords a species a significant evolutionary advantage. Therefore it is not surprising that plants have evolved a prodigious spectrum of strategies enabling their seeds to travel with animals. They ride on the plumage, skin, or fur of birds and mammals, or come wrapped in delicious fruit pulp so that they can stow away in the mouth or gut of an animal.

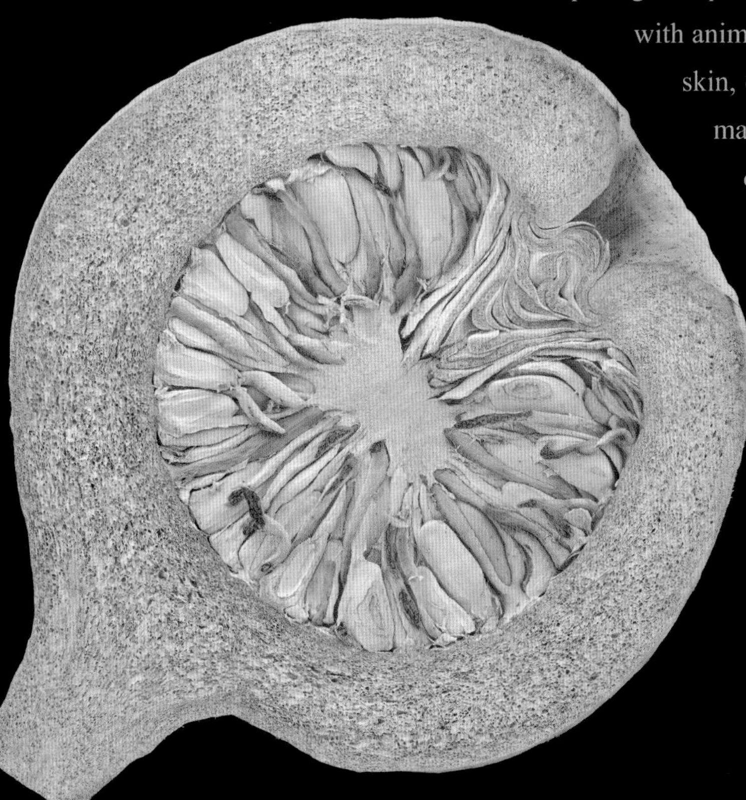

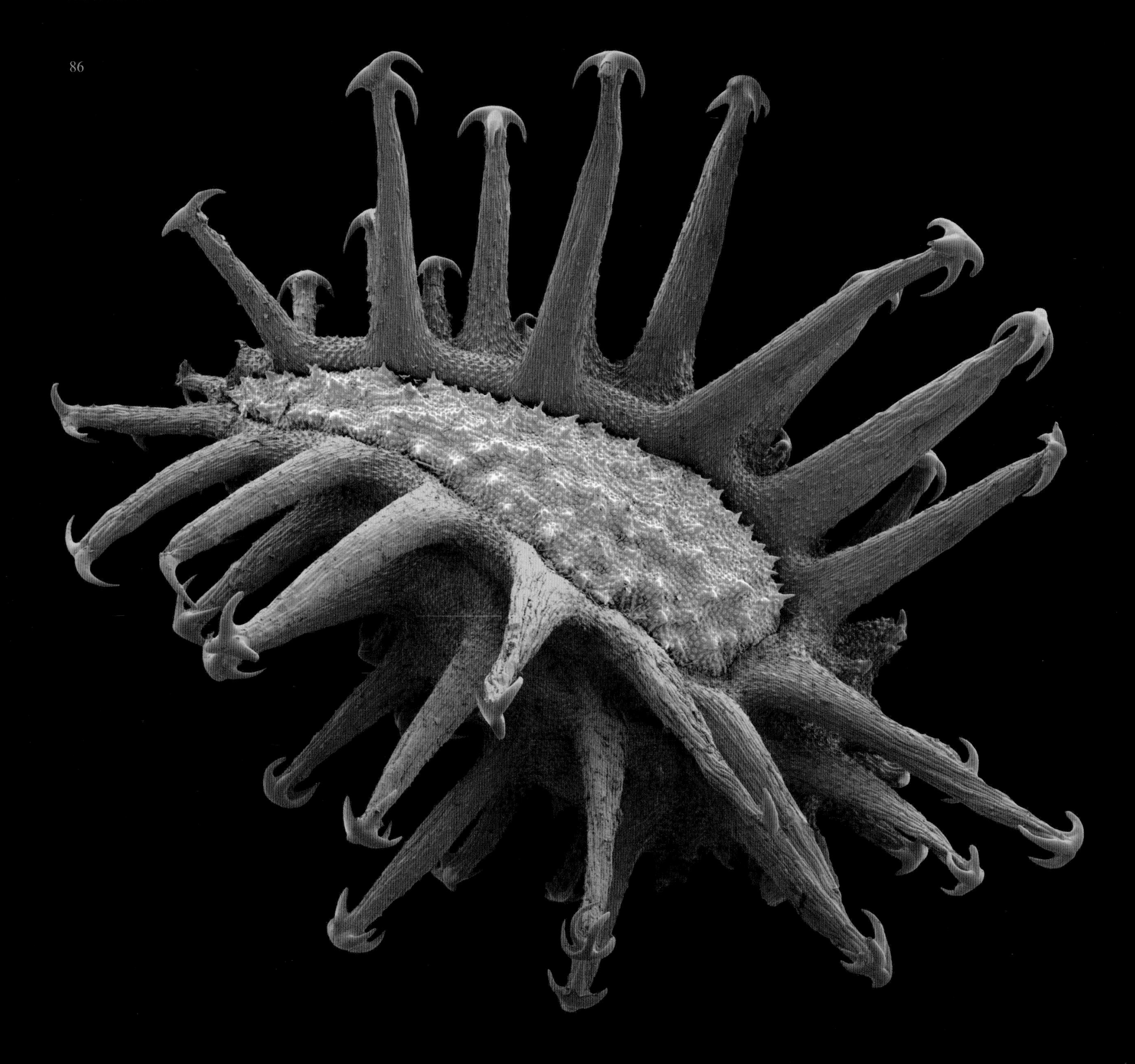

TENACIOUS HITCHHIKERS

Epizoochory or hitching a ride on an animal is a cost-effective way of travelling. and does not necessarily need special adaptations. Small diaspores without any modifications towards any particular dispersal mechanism often travel as stowaways in mud sticking to the feet or feathers of waterfowl or other birds and animals. Chance dispersals also occur when grazing animals ingest tiny seeds as they forage. The diaspores of many strategically-placed, low-growing plants are specifically modified to attach themselves to passing animals. Unlike fleshy fruits and seeds, adhesive diaspores do not offer any edible rewards to lure potential dispersers into picking them up, which means that dispersal happens by chance when an animal unwittingly 'becomes attached' to a seed. In addition to being physiologically cheap, epizoochory has another great advantage. Unlike fleshy diaspores, the dispersal distance of adhesive diaspores is not limited by factors like gut-retention times. Most sticky hitchhikers drop off by themselves but if they do not, they may travel long distances before they are groomed away or the animal moults or dies. Typical adaptations indicating epizoochorous dispersal are diaspores covered in hooks, barbs, spines or sticky substances. Examples are often found on socks and trousers after a country walk in late summer or autumn. Among the most frequent and tenacious burrs in our temperate climate are the nutlets of Stickywilly (*Galium aparine*, Rubiaceae), Hound's-tongues (*Cynoglossum* species, Boraginaceae), Wild Carrot (*Daucus carota*, Apiaceae) and Stickseeds (*Hackelia* species, Boraginaceae), and also the fruit burrs of Sticklewort (*Agrimonia eupatoria*, Rosaceae) and the much larger burrs of Burdock (*Arctium lappa*, Asteraceae). The underlying principle of adherence is simple, consisting of small hooks that readily become entangled with the mammalian fur, or tiny loops of fibre in clothes. It was the microscopic structure of these diaspores that, in the 1950s, inspired the Swiss electrical engineer George de Mestral to develop the hook and loop fastener that we know as Velcro®. However, the hook-and-loop principle is not the only way diaspores attach themselves to animals. Plants have also developed quite sadistic means to achieve seed dispersal.

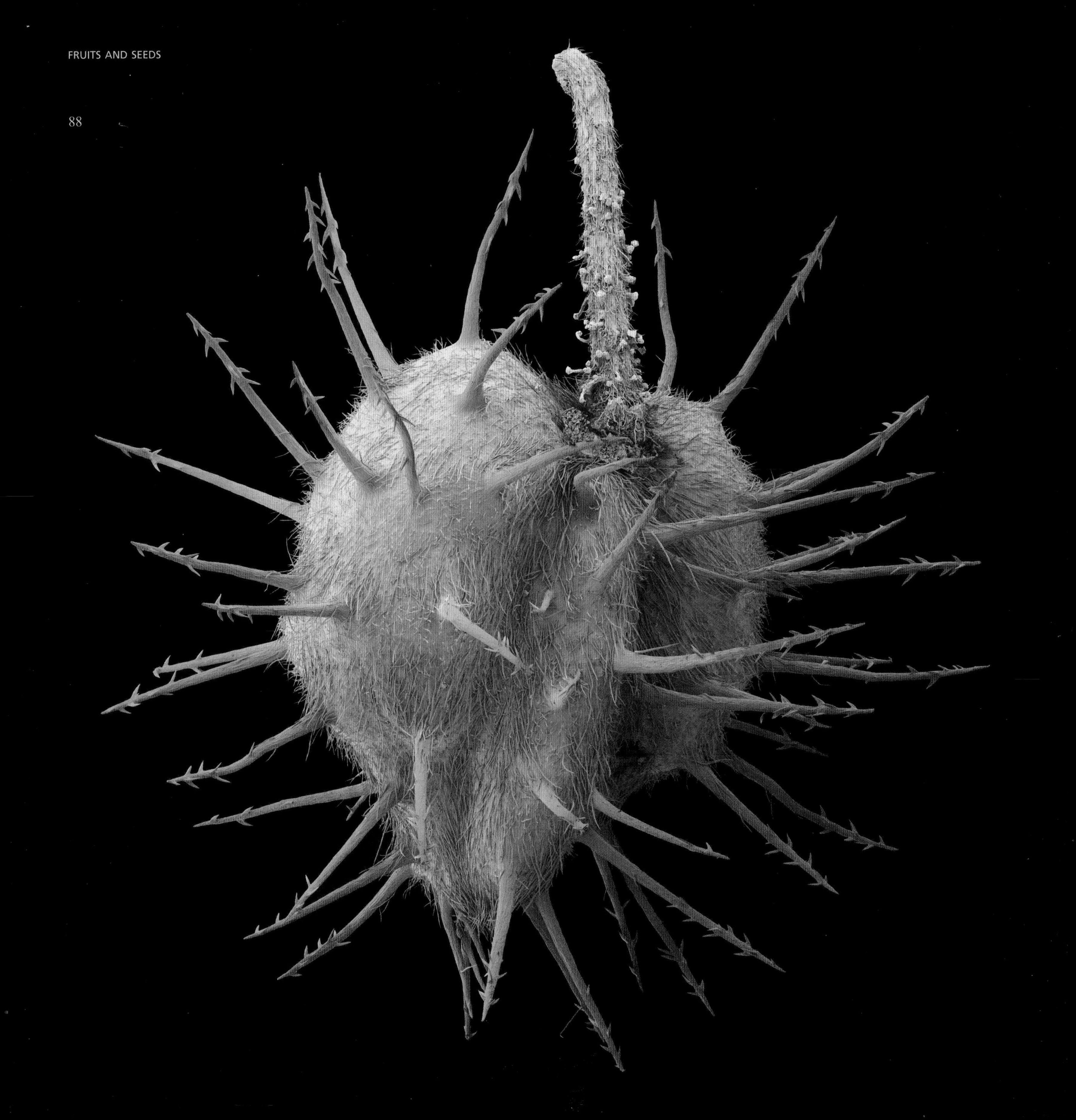

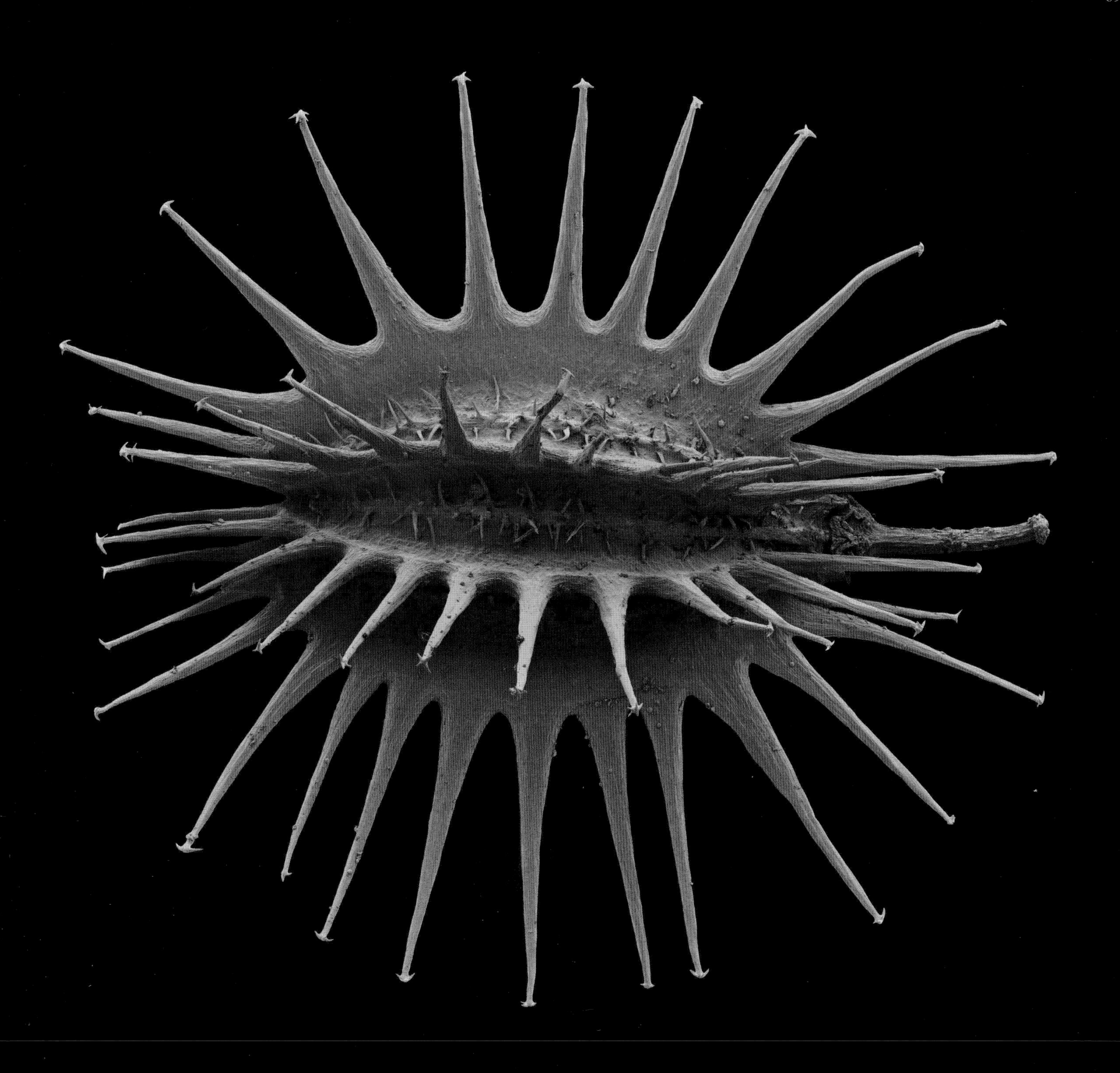

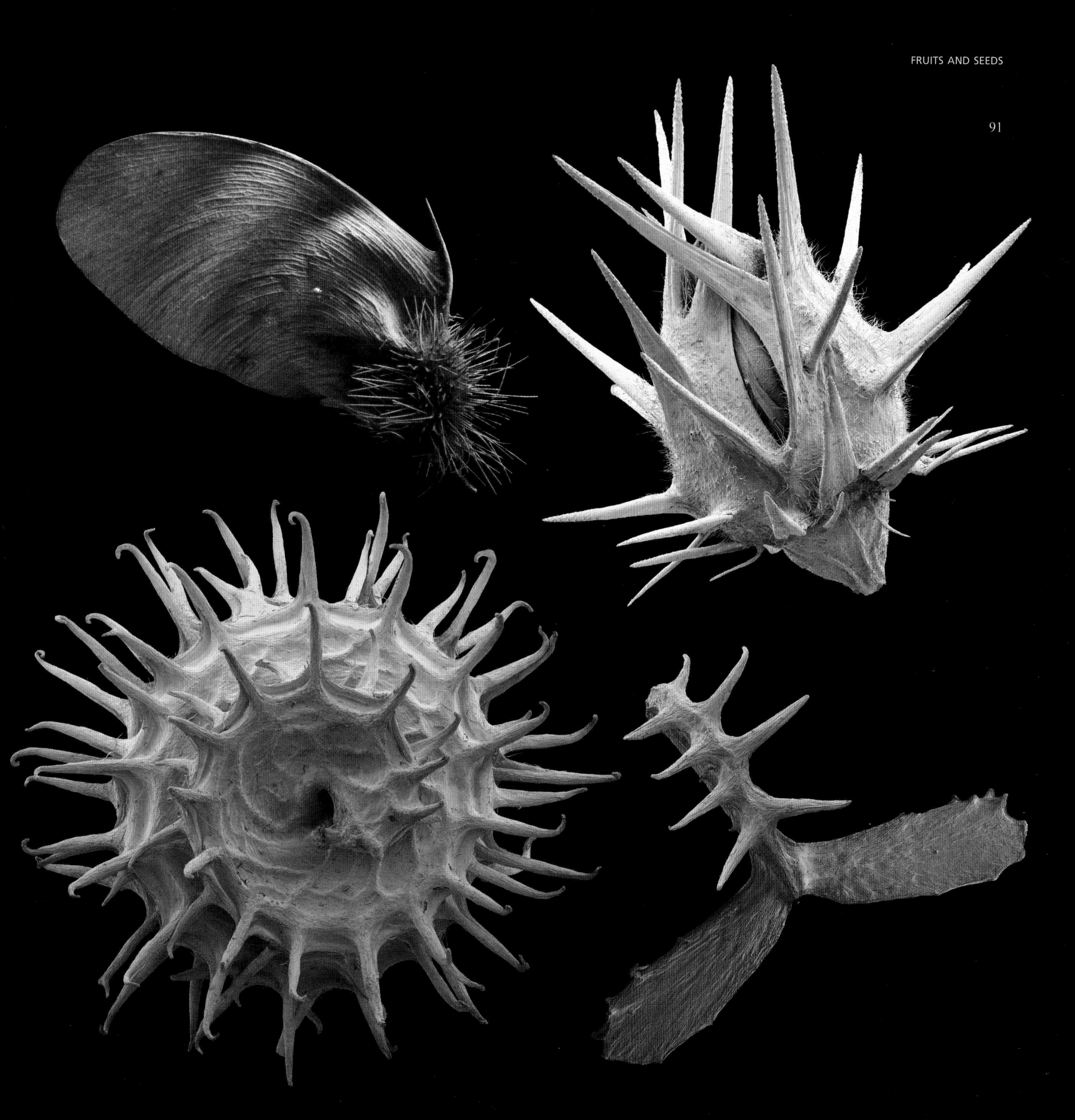

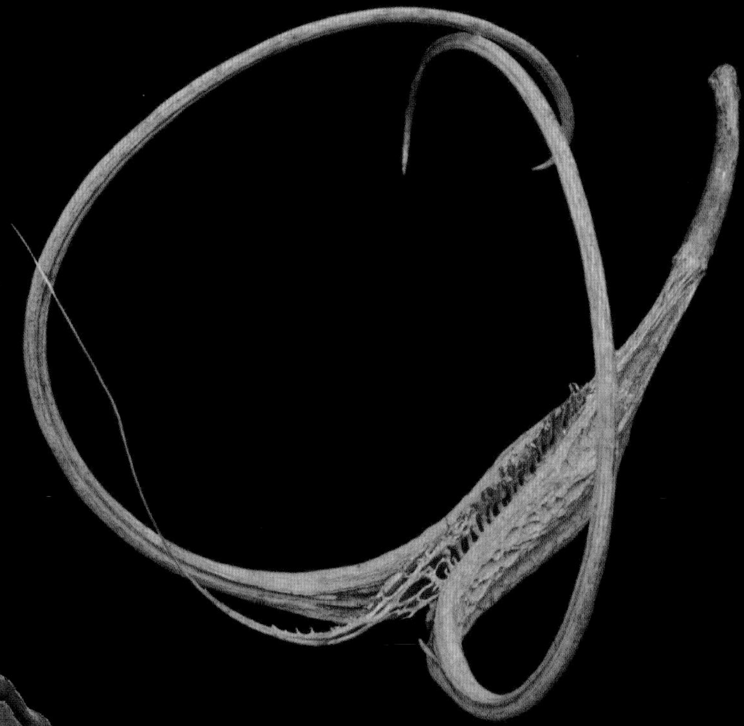

CALTROPS, DEVIL'S CLAWS AND OTHER SADISTIC FRUITS

Ferociously spiny diaspores that developed to bite into flesh occur in a variety of unrelated families. From the warmer parts of Europe, Africa and Asia there is the puncture vine (*Tribulus terrestris*, Zygophyllaceae). To some the plant is better known as Devil's Thorn or Caltrop, because of its diabolically insidious diaspores. As the *schizocarpic fruits* of the Puncture Vine mature they split into five indehiscent nutlets. Each nutlet is armed with two large and several smaller spines. Whichever way up they land, some of the spines point upwards like some medieval *caltrop* poised to penetrate the skin of an animal or the soles of shoes. In the Hungarian plains these prickly hitchhikers cause sheep farmers considerable problems by inflicting wounds that fester and make it difficult for an animal to walk.

The largest, most notorious burrs, known as Devil's Claws, are encountered in the tropical and subtropical semi-deserts, savannahs and grasslands of America, Africa and Madagascar. The New World Devil's Claws belong to the genus *Proboscidea* (notably *Proboscidea louisianica*) and their smaller relative *Martynia annua*, both members of the Unicorn family (Martyniaceae). In South America their carnivorous brethren from the genus *Ibicella* produce similar Devil's Claws, or Unicorn fruits as they are sometimes called (for example, *Ibicella lutea*). The immature green fruits look harmless. Their true nature is only revealed when the ripe fruits shed their outer fleshy cover and expose their highly ornate endocarps. At the tip of each endocarp is a beak that splits down the middle to produce a pair of sharply pointed, recurved spurs. Once the ferocious spurs have unfolded the fruit is ready to cling to fur or hoof and even bore into skin.

Old World Devil's Claws belong to the Sesame family (Pedaliaceae), close relatives of the Martyniaceae. The burrs of the Malagasy genus *Uncarina* look like miniature sea mines with long, sharply barbed spines. But in terms of cruelty, nothing can beat the fruits of the grapple plant, *Harpagophytum procumbens* from southern Africa. The tardily dehiscent woody capsules bear numerous thick, sharply pointed, barbed hooks that inflict serious wounds on anyone who steps on them.

REWARD RATHER THAN PUNISHMENT

Instead of maliciously exploiting the ability of animals to move , the majority of *zoochorous* plants have developed mutually beneficial relationships with their dispersal partners: in exchange for their courier services the animals receive an edible reward.

SMALL REWARDS FOR LITTLE HELPERS

On close examination, it can be observed that the seeds of many plants, especially in dry habitats, have a small yellowish-white oily nodule. In 1906 the Swiss biologist Rutger Sernander described the strategy that lies behind these strange appendages and called it *myrmecochory*. (Greek *myrmex* = 'ant' and *choreo* = 'to disperse'). He found that seeds bearing such an 'oil body' – or *elaiosome* as he called it in Greek – were irresistible to ants, which collect the seeds avidly and carry them to their nest. What triggers this stereotypical, seed-carrying behaviour is the presence of ricinolic acid in the elaiosome. As the result of millions of years of co-adaptation, myrmecochorous plants have evolved to produce the same unsaturated fatty acid in the tissue of their elaiosomes as that found in the secretions of the ants' larvae. Once the workers have hauled the seeds into the nest, they dismantle the nutritious appendage but leave the rest of the seed, which is well protected by a hard seed coat, unharmed. The tissue of the elaiosome, rich in fatty oil, sugars, proteins and vitamins, is not consumed by the ants themselves but used to feed their larvae. Once the fatty nodule has been removed, the seeds are discarded on a nest midden, which can be either underground or above ground. The organic substrate of such refuse piles is rich in nutrients and offers better conditions for seedling establishment and growth than the surrounding soil.

Obviously, ant-dispersed seeds have to be small to match the physical strength of their dispersers. Adaptation to ant-dispersal is common among herbaceous plants in temperate deciduous forests in Europe and North America. In dry habitats prone to frequent fires, such as the Australian heathlands and the fynbos in the Cape region of South Africa, myrmecochory plays an even more important role. Storage in an underground formicary greatly increases the chance of escaping destruction by fire as well as consumption by seed-eaters such as rodents. Ant-attracting seed appendages are found in many different plant families including the spurge family (Euphorbiaceae).

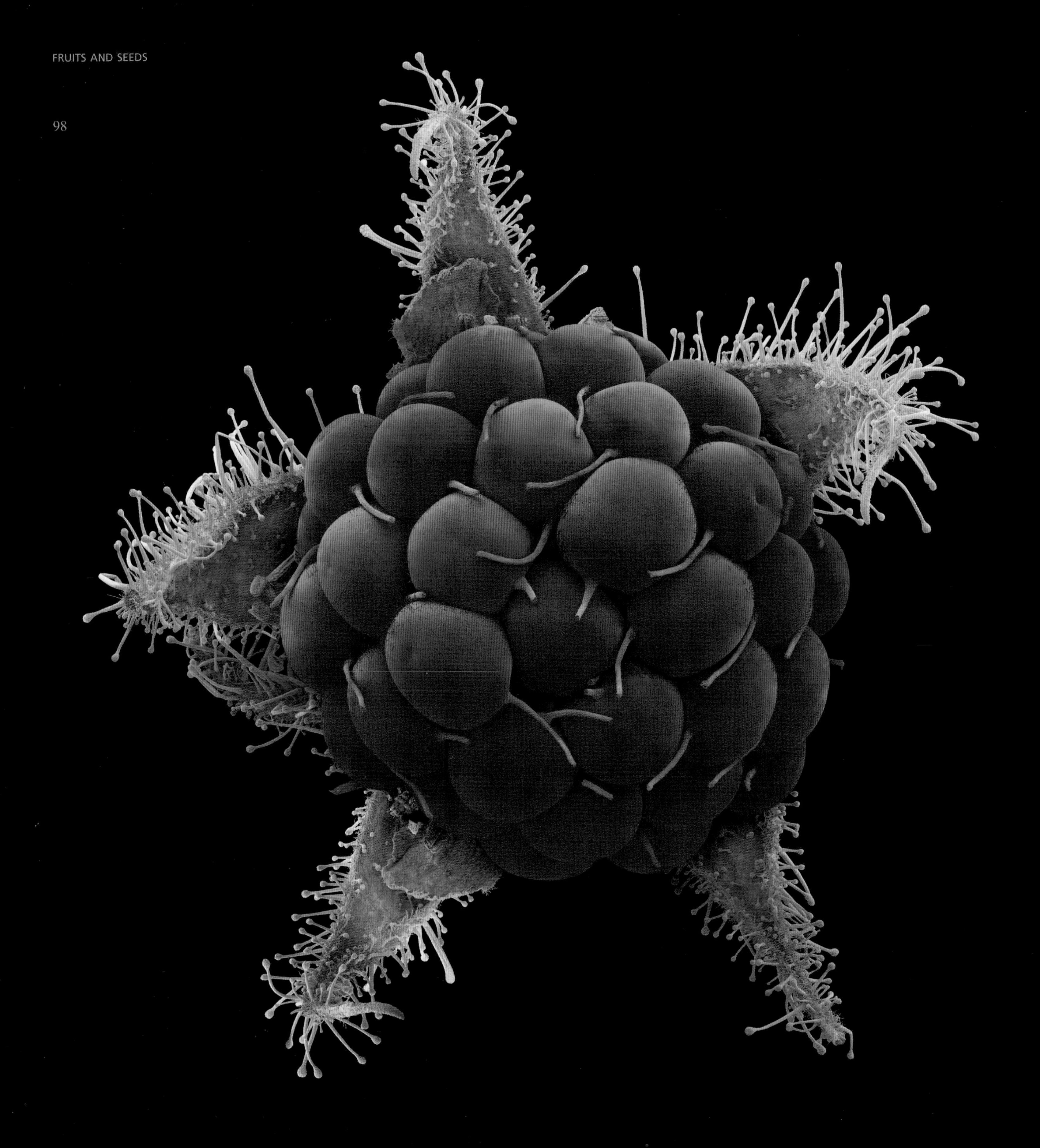

JUICY TEMPTATIONS

The best proof that edible rewards provide a convincing attraction for animal dispersers is our own fondness for fruit. Behind every sweet and juicy fruit we enjoy lies the ruthless intention of a plant to disperse its seeds. The sweet pulp of a fruit is nothing more than bait, designed to manipulate a potential disperser into eating the tasty morsel and swallowing it with its seeds. After its meal the animal moves on, unintentionally giving the seeds a lift. As the hours pass the meal is digested and eventually the stowaways disembark with the faeces. With luck, some of the seeds will land in a suitable place to germinate, well away from the shadow of the mother plant. This form of dispersal is called *endozoochory*, 'dispersal inside the animal'.

Among vertebrates, the most important dispersers are birds and mammals, especially in our temperate climate. In the tropics fruit birds, fruit bats and monkeys are the most important dispersers, together with some fish and reptiles but these play only a minor role.

While fruits are ripening they are inconspicuous, rather hard, and have no smell; they are sour at best; at worst they are poisonous. Altogether, they make sure they are as unpalatable as possible a long as the seeds are immature. As soon as the seeds are ready to be dispersed, the fruit sends out signals that promise a safe, nutritious reward. The nature of the signals depends on the kind of animal they want to attract. Birds have excellent colour vision but a poorly developed sense of smell. Therefore, diaspores adapted to bird-dispersal (*ornithochory*) have no scent. Instead, they catch the birds' attention by changing their colour from green to something much more conspicuous. Red is the colour birds distinguish best against a green background, but purple, black and sometimes blue – or combinations of these (especially red and black) – are also found. A slightly different strategy is more likely to attract mammals, who rely more on their keen sense of smell than on their eyesight and many of them are nocturnal. Mammal-dispersed fruits are therefore often (but not always) dull in colour (brown or green) and emit a strong aromatic scent when they are ripe. Apples, pears, medlars, quinces, citrus fruits, mangoes, papayas, passion fruits, melons, bananas, pineapples, jackfruits, breadfruits and figs are examples of fruits that target mammals such as rodents, bats, bears, apes, monkeys, and even elephants and rhinos, as dispersers for their seeds.

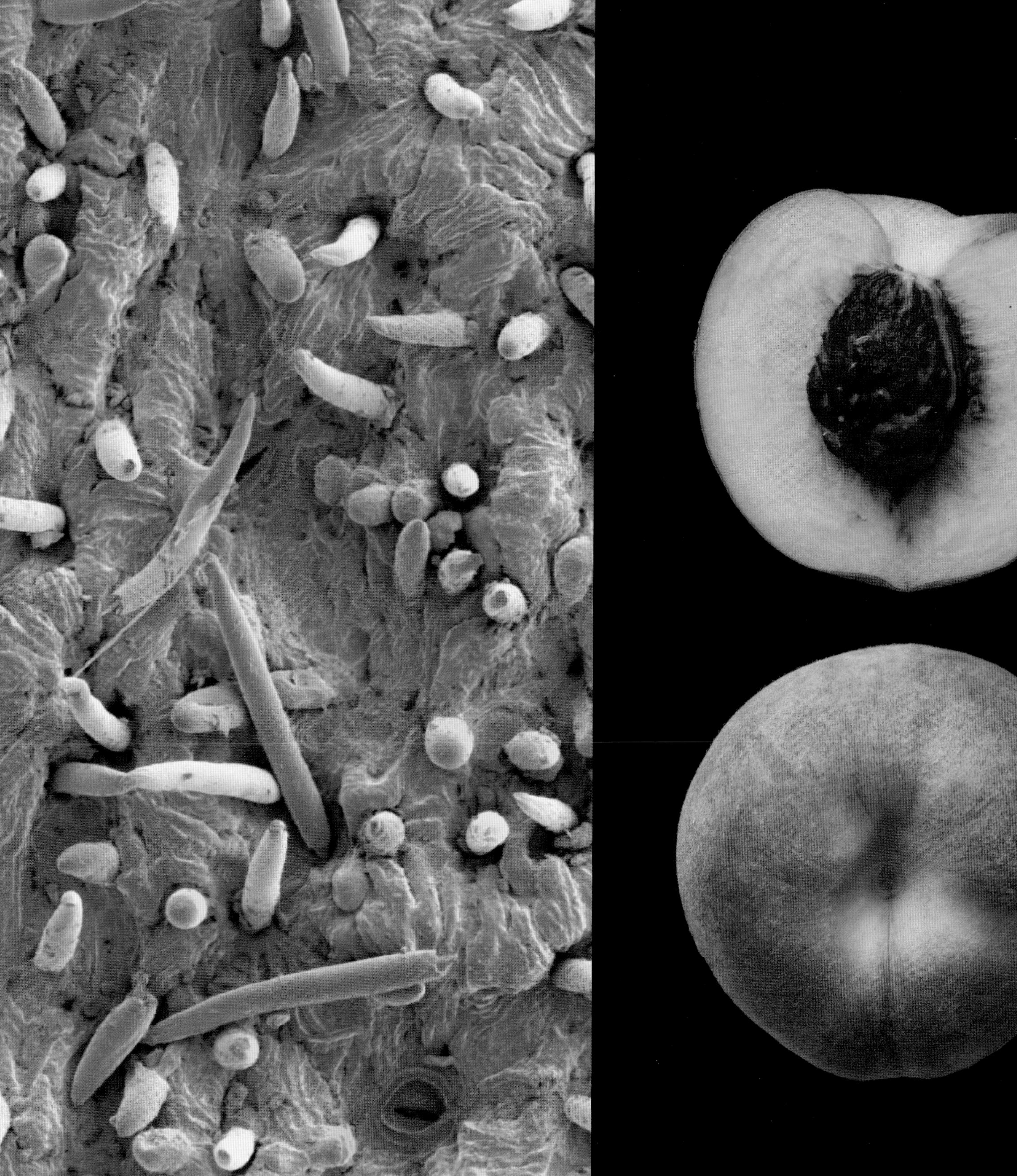

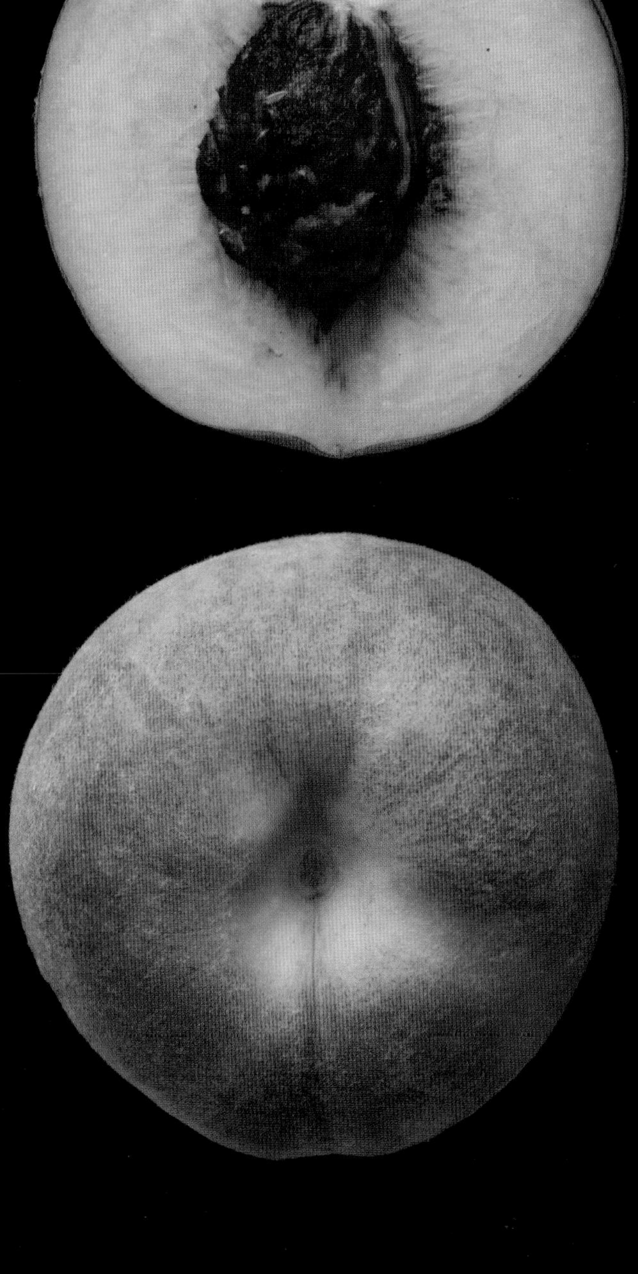

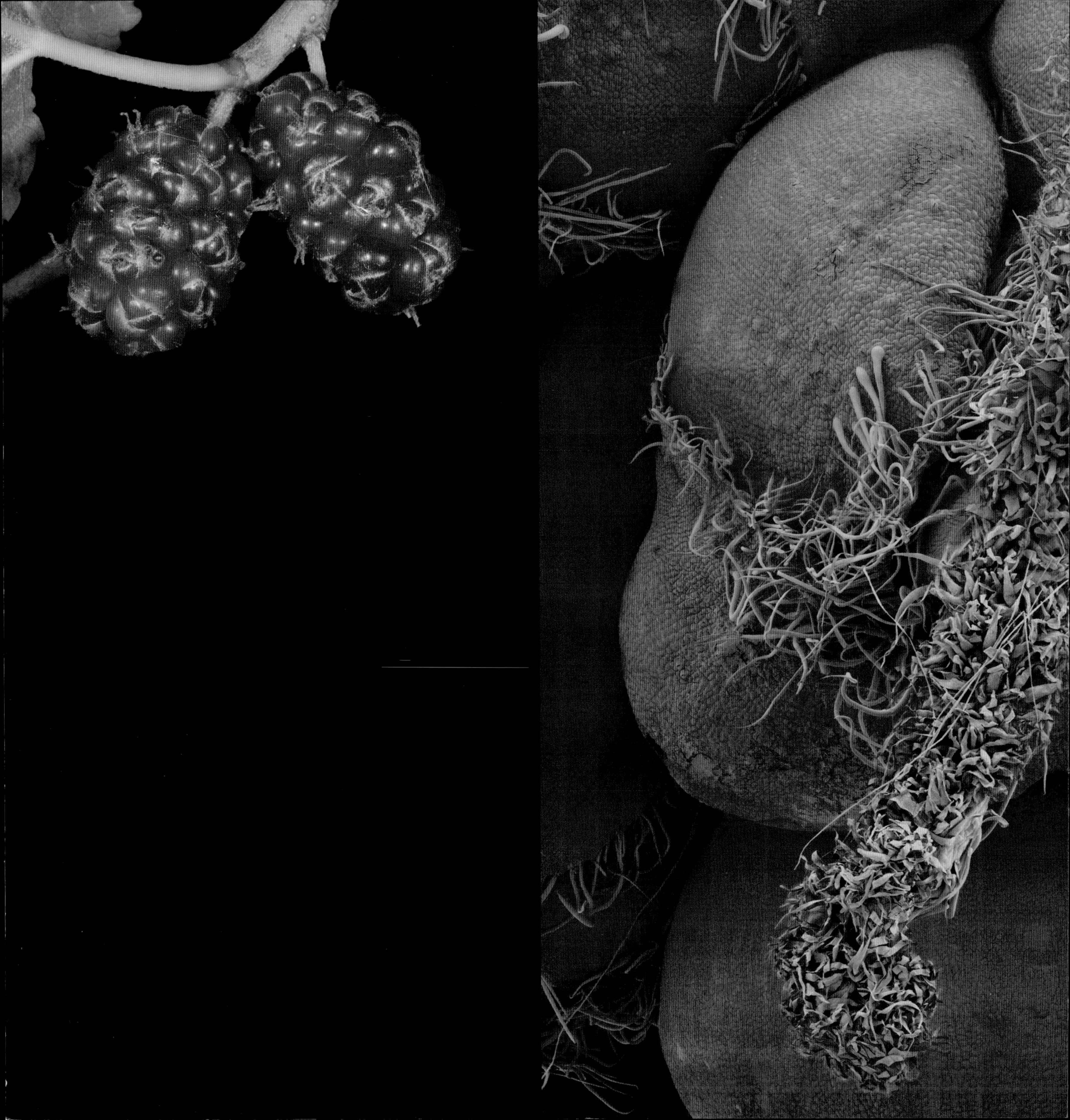

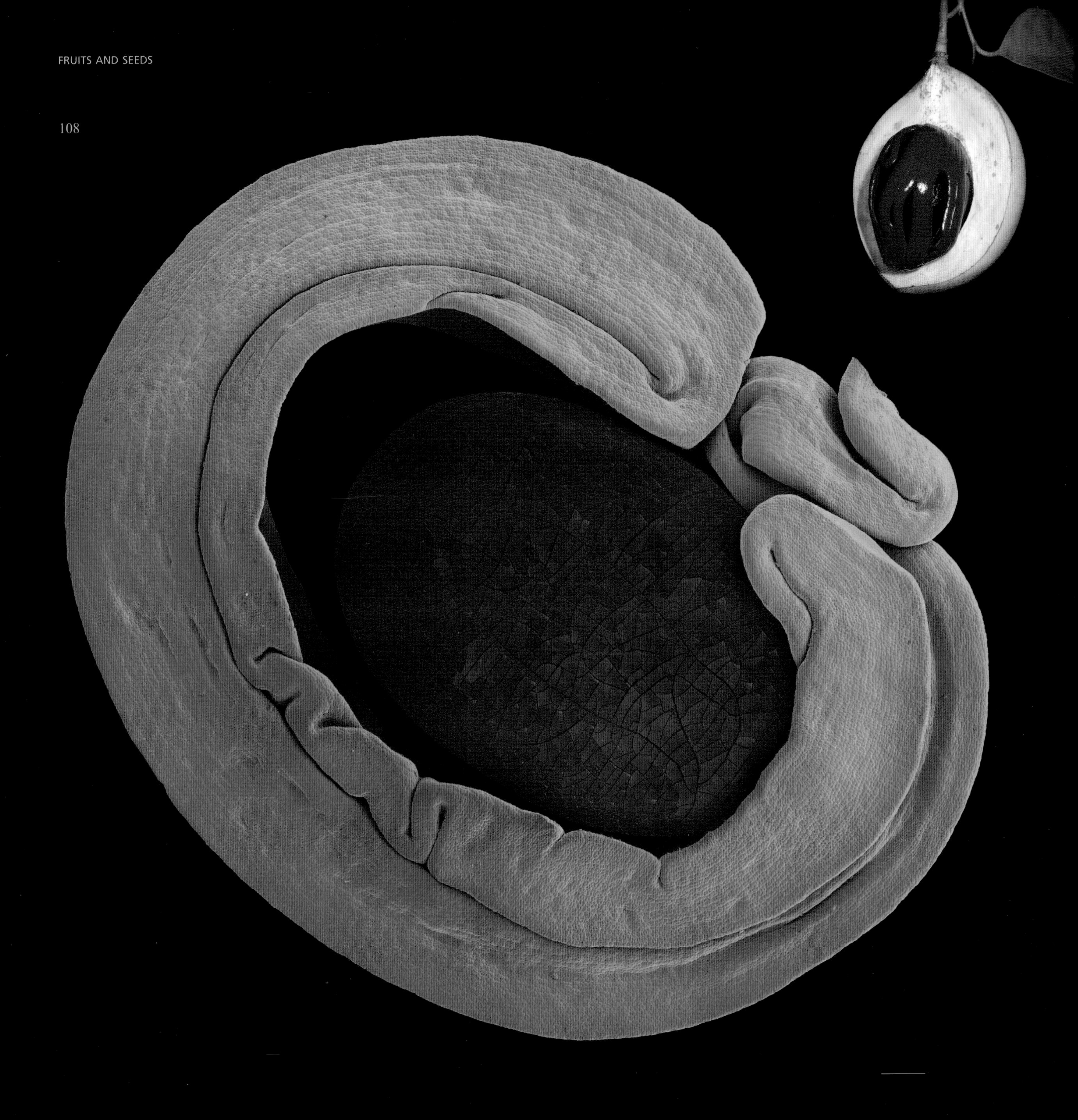

COLOURFUL APPENDAGES

The edible reward of animal-dispersed fruits is typically a fleshy, edible fruit wall. Large seeds may bear appendages, called *arils*, to attract birds. Stark contrasts are an important lure for avian dispersers, and arils play their part. *Euonymus europaeus* (Celastraceae), the Spindle Tree, is one of the few north-temperate examples with spectacularly coloured fruits and arillate seeds. The bright red, loculicidal capsules open to expose three or four seeds wrapped in a deep orange aril. Once the pendulous fruits are fully open the seeds drop out and dangle from short 'umbilical cords' (*funicles*) adding movement to the display. But as usual the tropics boast much larger, far more colourful examples. Black seeds with white arils against red fruit walls appear to be a successful dispersal device, which has evolved independently in various plants including the Manila Tamarind (*Pithecellobium dulce*, Leguminosae) and the Chaquiro (*Pithecellobium excelsum*) from tropical America. Another common version of the bird-dispersal syndrome is the display of black seeds with orange or red appendages against the bright background of the fruit wall, a pattern found in the loculicidal capsules of the bird-of-paradise flower (*Strelitzia reginae,* Strelitziaceae) from South Africa; its seeds have a curious aril resembling a shaggy orange wig. The same syndrome occurs in bird-dispersed legumes including the African Mahogany (*Afzelia africana*) and the Australian Coastal Wattle (*Acacia cyclops*). The New Zealand Titoki Tree has black seeds wrapped in fleshy red arils which emerge suddenly from inconspicuous, greenish-brown fruits.

The most precious red aril of all is inside a rather humble fruit. When the thick-walled fruit of the nutmeg (*Myristica fragrans*, Myristicaceae) – initially green and later pale yellow to light brown – is split down the middle, a single large seed with a spectacular, lacy, crimson-red aril is revealed. The seed and the aril, known as nutmeg and mace, were the most precious commodities of the spice trade for hundreds of years. They have a similar, highly aromatic taste but the aril is considered more delicate. Their natural dispersers are birds. In Indonesia, Pigeons from the genus *Ducula* and Hornbills (family Bucerotidae) are probably their most important natural dispersers.

FRAUDSTERS OF THE PLANT KINGDOM

At the end of our journey into the bizarre and incredible world of plants, one final, rather sinister side of plants to be examined is their fraudulence. Wherever there are thriving relationships between two species to the benefit of both partners, there will be cheats who, without compensation, try to take advantage by giving nothing in return for the service or the benefit they receive. This cost-saving strategy is not only a sad reflection on human society but a general pattern in nature. Plants are no exception: saving material and energy affords an evolutionary advantage. Some frugivorous birds and monkeys become pulp thieves: they eat only the juicy parts of a fruit and drop the seeds under the parent tree. On the other side, some plants have evolved strategies to trick animals into swallowing their seeds without providing food in exchange. Grasses deceive large herbivores by hiding their small, dry fruits among their leaves, a ploy for which the ecologist Daniel Janzen (1984) coined the phrase 'the foliage is the fruit'. Other plants are openly deceitful and offer fruits or seeds with contrasting colouration that imitates fleshy ornithochorous diaspores such as berries, drupes or arillate seeds. They appear to offer food, but in reality there is no edible reward that is nutritious for the animal but energy-expensive for the plant.

Although the concept of fruit mimicry remains controversial, experiments have shown that at least some naïve frugivorous birds mistake the deceptive seeds of mimetic fruits for fleshy diaspores and eat them. Examples of mimetic fruits are rare and belong mainly to the bean family (Leguminosae) with occasional suspects in other families such as the Sapindaceae (for example, *Harpullia* species). A common strategy among legumes is to offer inedible black, red, or contrasting red and black seeds against the beige or pale yellow to deep orange-red background of the equally inedible inner carpel wall. The

Snow Wood (*Pararchidendron pruinosum*), a small Australasian rain-forest tree is also called 'monkey's earrings' because of its eye-catching twisted pods that boast shiny black seeds against the gaudy red background of the inner fruit wall. The unusual fruits of New Zealand's North Island Broom (*Carmichaelia aligera*) also raise suspicions of deceit. Once the fruit walls have dropped off, the shiny red seeds with occasional black spots remain on permanent display surrounded by the contrasting black frame of the fruit.

Despite their attractive, 'juicy' appearance, the fruits and seeds of the Snow Wood, the New Zealand Broom and their fellow fraudsters are hard and dry. This renders them useless to fruit-eating birds but for enthusiasts of botanical jewellery they are treasures. Among the favourites are the pure red seeds of Coral Trees (members of the genus *Erythrina*) and others such as the Red Bead Tree (*Adenanthera pavonina*) from south-east Asia and Australia, the Texas Mescal Bean (*Sophora secundiflora*), a native of the south-western United States and Mexico, and the Panamanian *Ormosia cruenta*. Even more eagerly sought are the bicoloured red and black seeds of the pantropical Rosary Pea (*Abrus precatorius,* also called Crab's Eye, Jequirity Bean or Paternoster Bean), the Jumby Bean (*Ormosia monosperma*) from South America and the Caribbean, and the American Rosary Snoutbean (*Rhynchosia precatoria*).

As makers of botanical jewellery can confirm, there are not very many species with hard, shiny but colourful seeds. Indeed, as with most scams, offering flashy but inedible seeds to unwitting hungry animals will work only as long as the fraudsters are in the minority. Too many mimetic seeds would frustrate *bona fide* dispersers and force them to find a more reliable food source. In the end, they would take neither the deceptive nor the edible seed, and all parties would lose out. As with all matters related to the biotic world, evolution by natural selection maintains a careful balance between all living things.

THE EVERLASTING BEAUTY OF PLANTS

We hope that we have inspired an increased love and enthusiasm for plants and the creatures that depend on them, including ourselves. The kaleidoscope of extraordinary images in this book are a constant reminder that plants are not simply useful but, in their remarkably sophisticated survival strategies, are also extraordinarily beautiful. Above all, plants are alive and struggling to survive like the rest of us, which makes their beauty even more mysterious and awesome.

This book is a lively, educational celebration of the bizarre and incredible life of plants, but none of us can ignore the serious problems currently threatening to change or destroy nearly every aspect of life on our planet. Overpopulation by man has caused massive and thoughtless destruction of natural habitats, triggering an ever-accelerating global extinction of many plant and animal species. Extinction of a species is for ever and takes away another complex piece of a vast and beautiful jigsaw puzzle, which is the result of millions of years of evolution. Tragically, on top of this, the extinction of a species also affects many other species with which it has shared a habitat for thousands if not millions of years. The web of life is an intricate network of multiple interdependences where the ripple effects radiating from each extinction are unpredictable. Each dying species leaves a rent in the web of life,and every time a species dies, the web of life is weakened a little more. Human beings were once a thread in the web of life but, from about ten thousand years ago, man began to exploit the Earth's resources, and in so doing began to destroy it by cutting more and more of its threads. We must constantly remind ourselves that it is the very diversity of the web of life that carries us. By destroying it we are, quite simply, sawing off the branch we are sitting on.

The fossil record tells us that life on Earth has already experienced five global mass extinctions. After each disaster the recovery of global biodiversity took between four and twenty millions years. In human eyes, this unimaginable time span means 'eternity' for modern humans like ourselves have existed for no longer than about two hundred thousand years. Therefore, if our environment is to have any chance of short-term recovery, we must act now.

We are becoming increasingly aware of our catastrophic impact on the environment. Perverse as it may seem, the emerging panic over the threats that overpopulation and climate change pose to our 'civilization', sends out a ray of hope. *Homo sapiens sapiens* ('Knowledgeable Man') may yet find a way to honour this name and let reason prevail over selfish instincts.

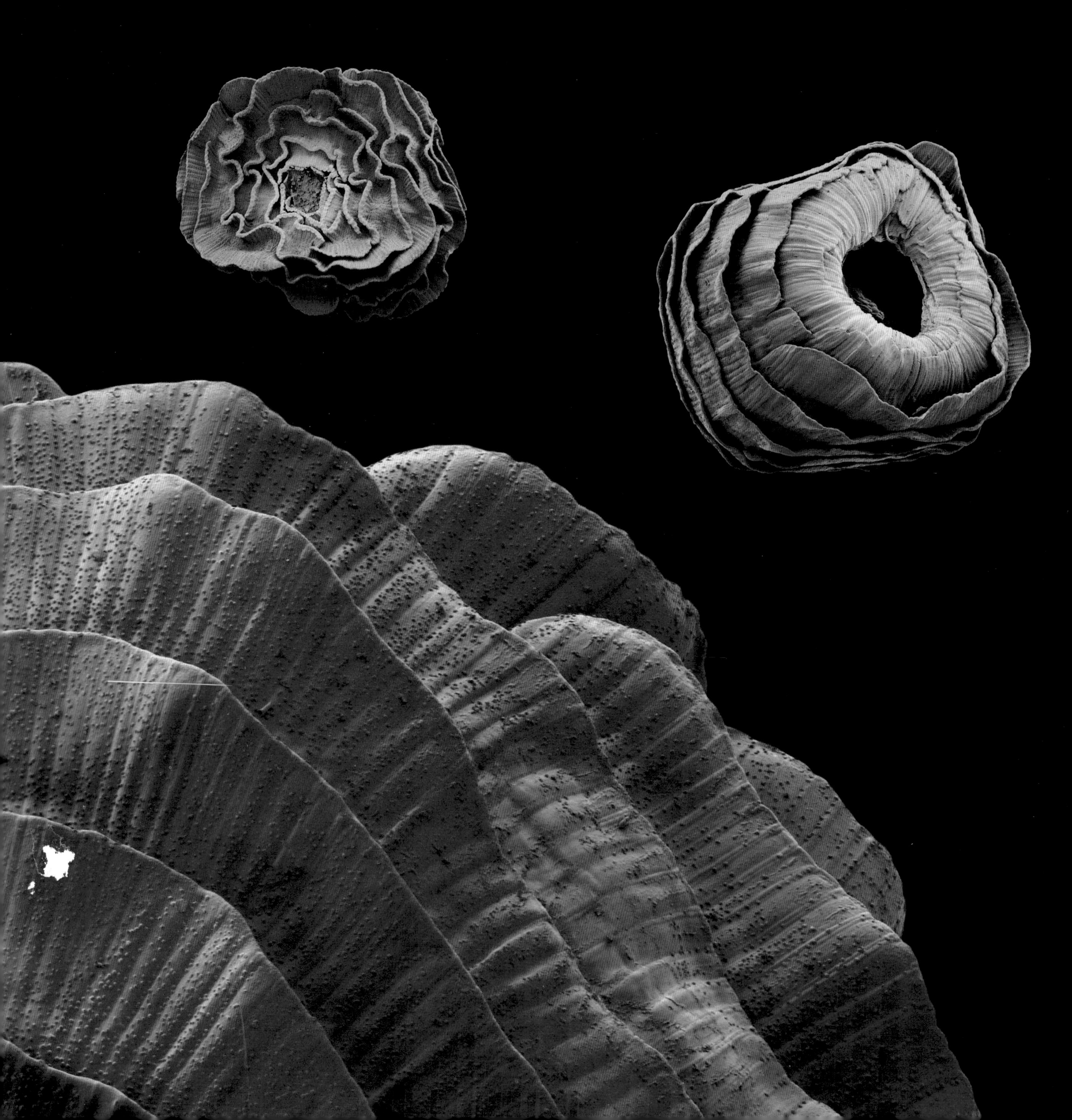

PHYTOPIA

With its seemingly endless array of colour and structure, the plant world has inspired generations of artists and illustrators, resulting in a spectacular wealth of paintings and illustrations that have served to inform and captivate many audiences across cultures and centuries. Approaches to portraying plants reflect the diversity of source material and the intention of the artist from the accuracy of botanical illustration for purposes of scientific record to an apparently limitless range of expressive imagination and interpretation. The development of photography in the 18th and the increasing sophistication of microscopes by the late 19th centuries revealed a new landscape of fantastic forms. However, in the 20th century, and particularly after World War II the tremendous advances that had been made in electron microscopy, originally for materials science, were discovered by the biologists. The light microscope was no longer the only method for exploring the 'invisible' in the natural world. Thus, for many years these highly specialised and enormously costly pieces of equipment were only available in a few centres of excellence to be used and shared by experimental scientists. Fertile opportunities for artistic intervention were not yet dreamed of. Now, with the remarkable developments in digital imaging over the past twenty years a common language has evolved that gives scope and impetus for extraordinary collaborations between the artistic and scientific communities. The spectacular images of pollen, seeds and fruit in this book are evidence of the manipulative hand of the artist exploiting some of the potential that these new technologies offer, where, by pushing the boundaries of scanning electron microscopy with colour manipulation, images of scientific integrity can also evoke a disturbing sense of hyperreality.

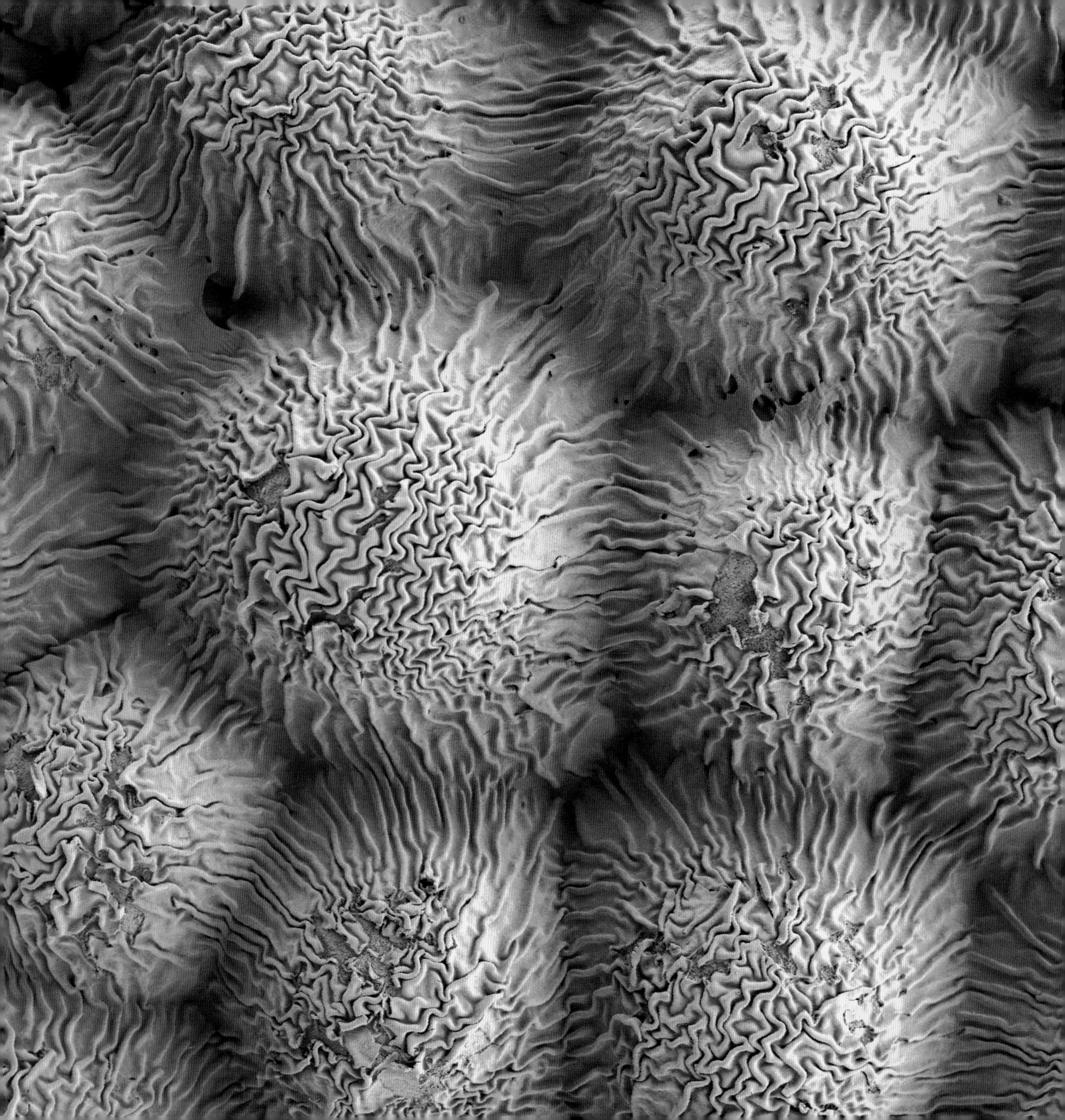

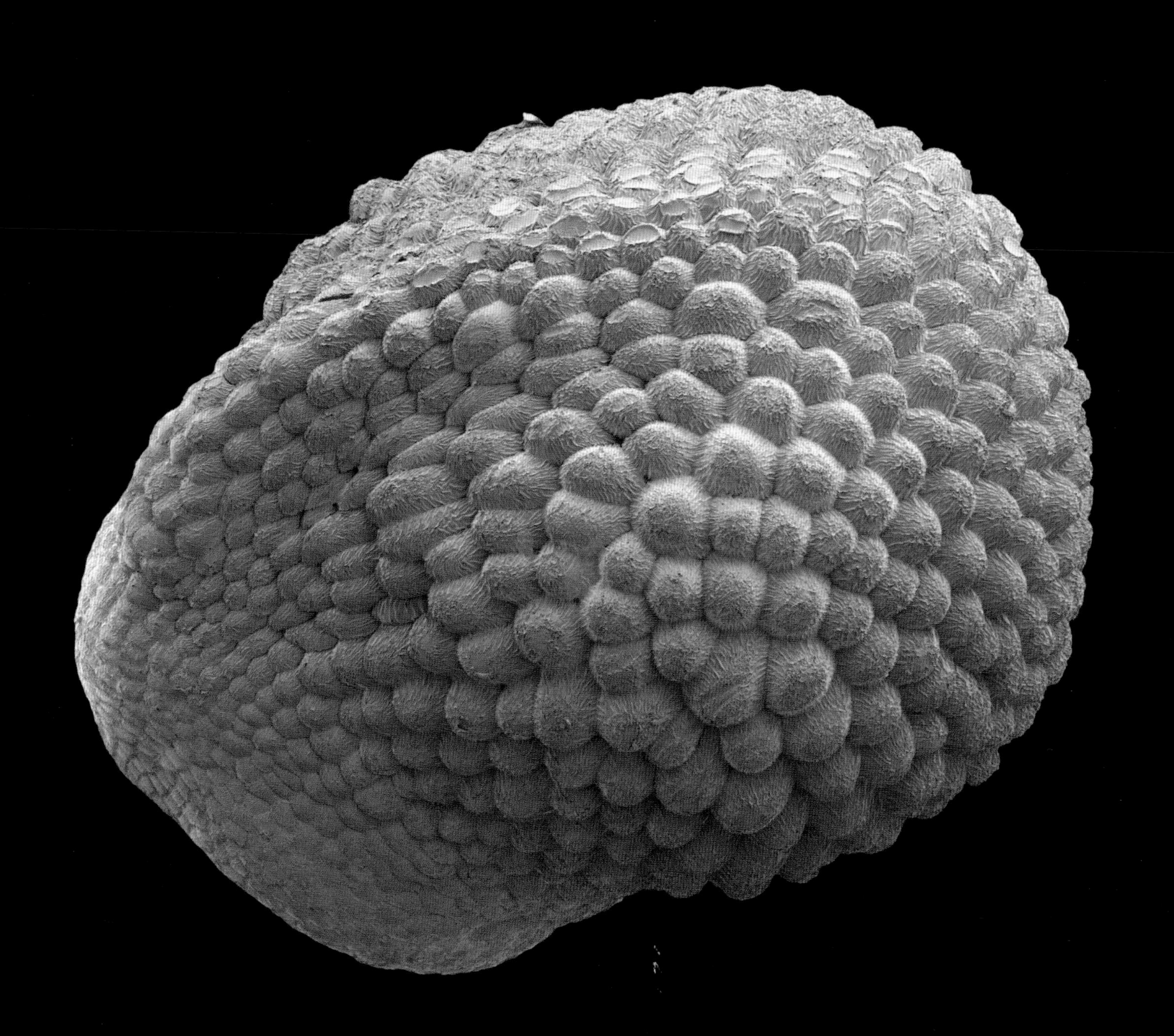

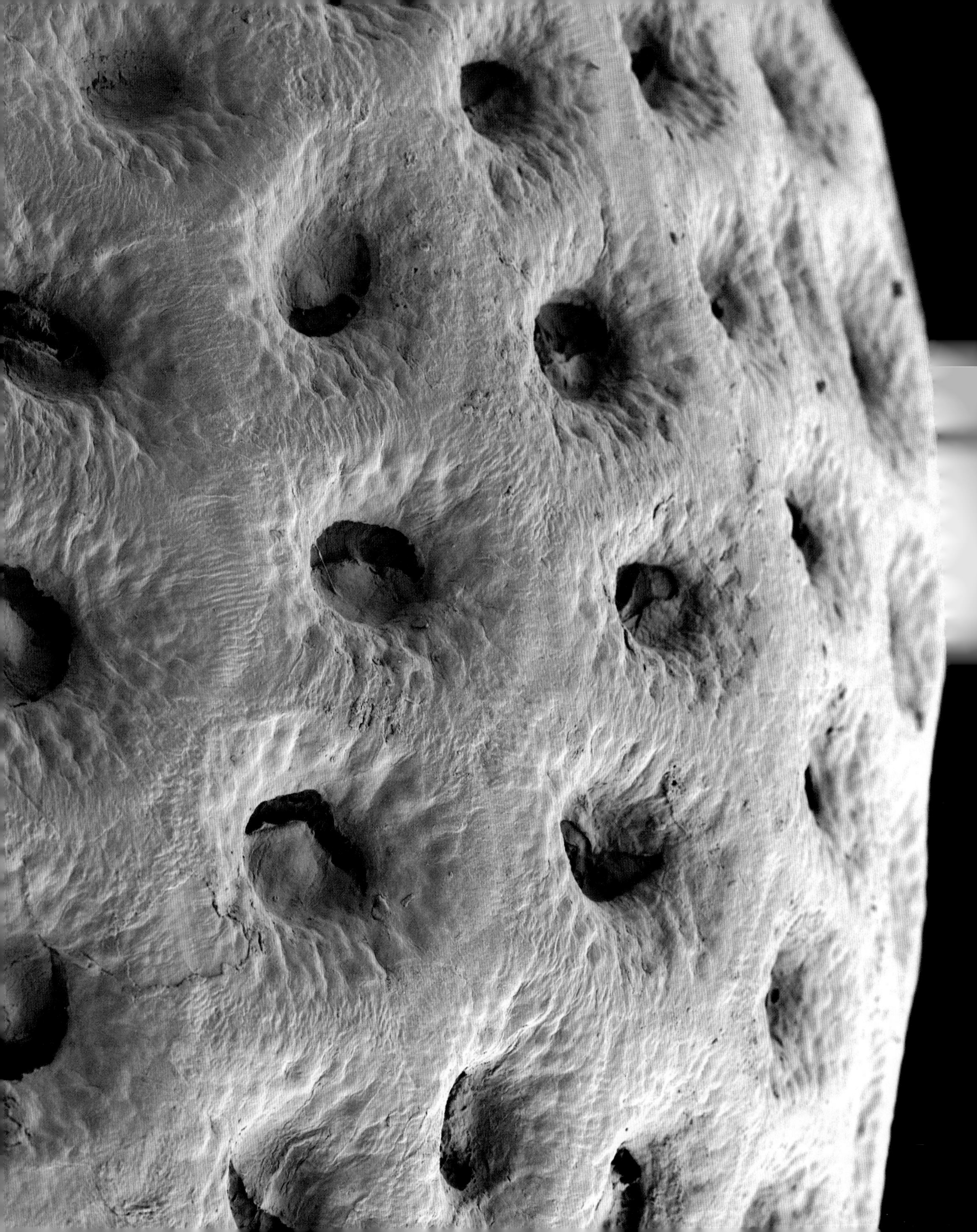

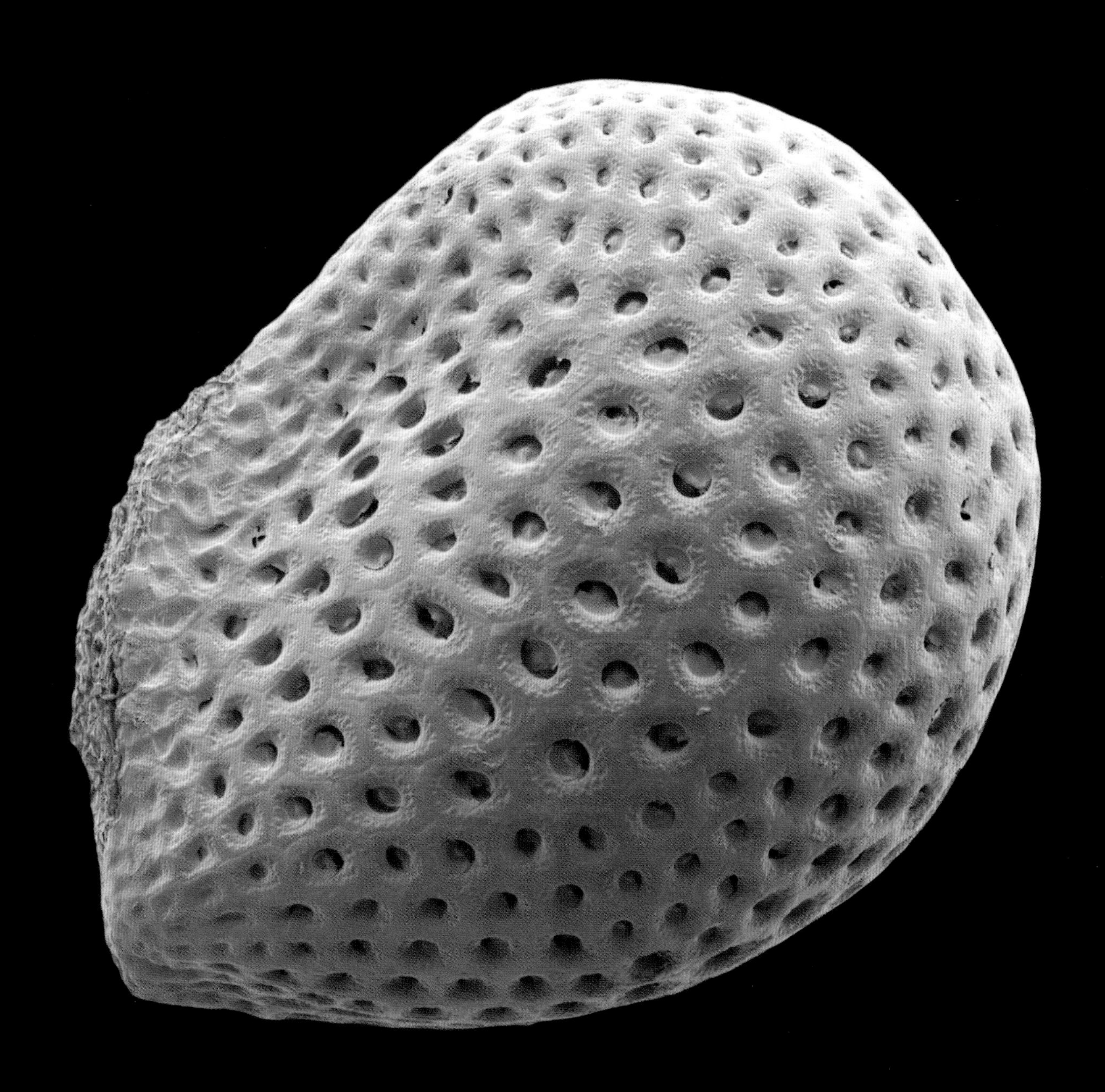

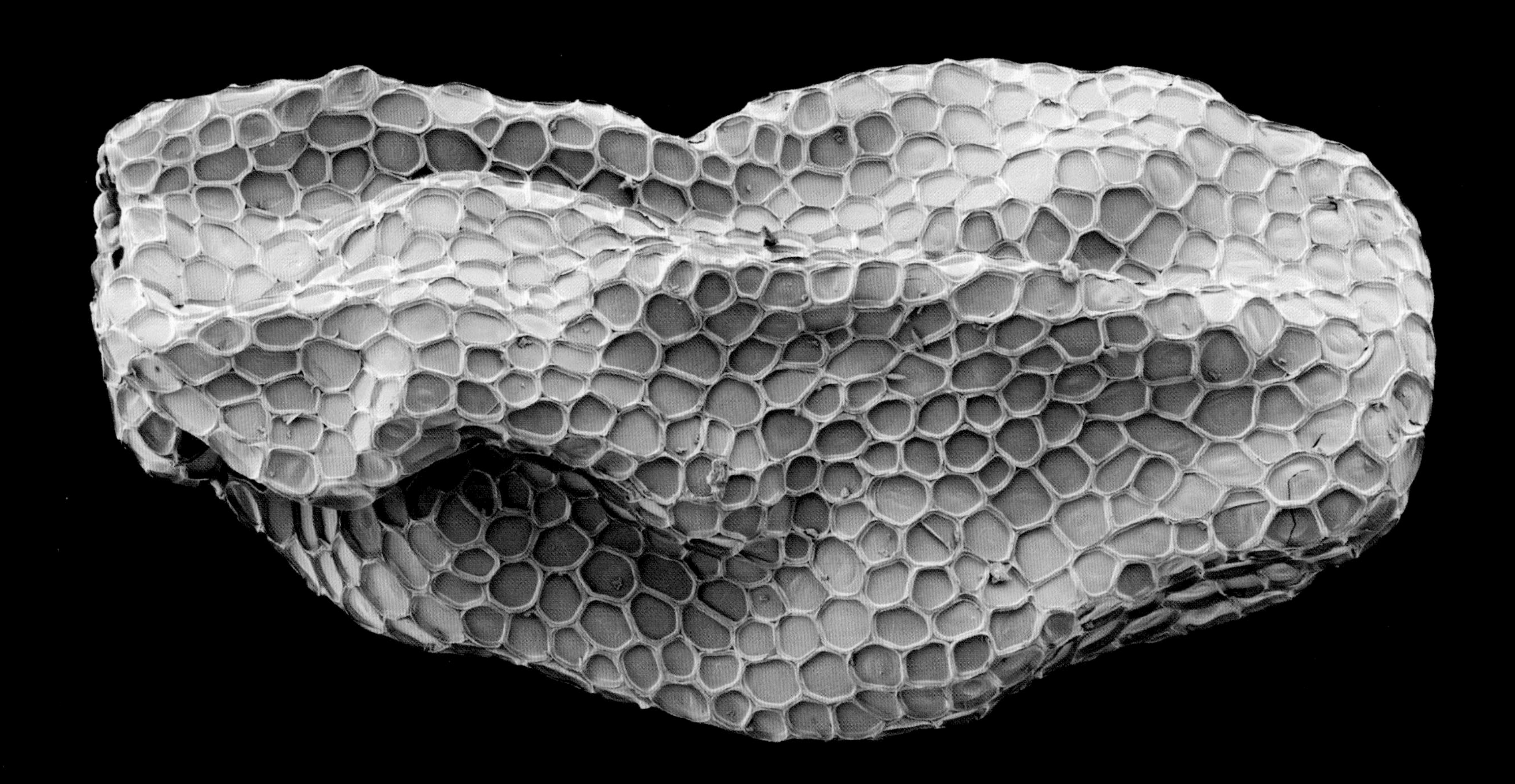

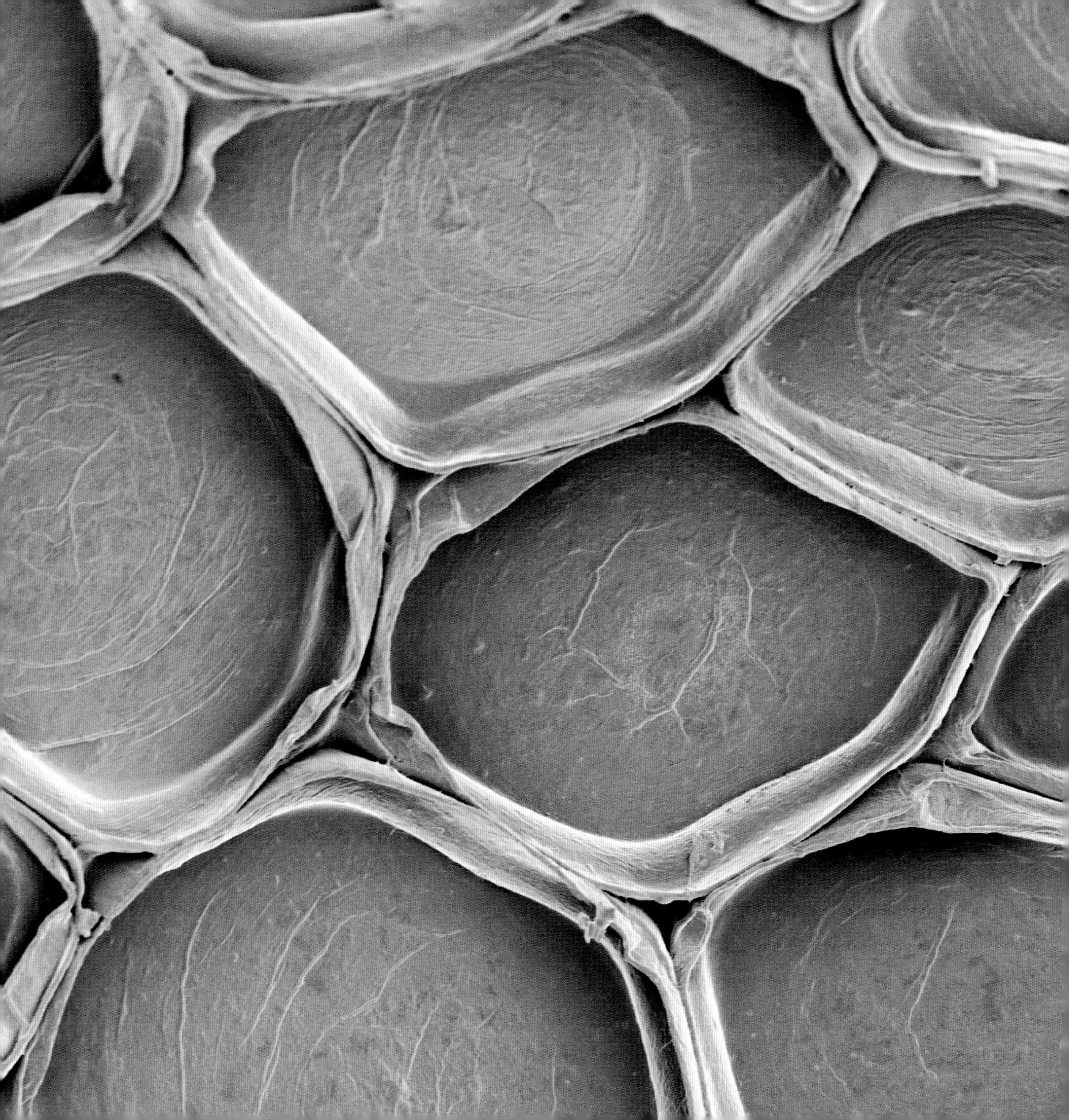

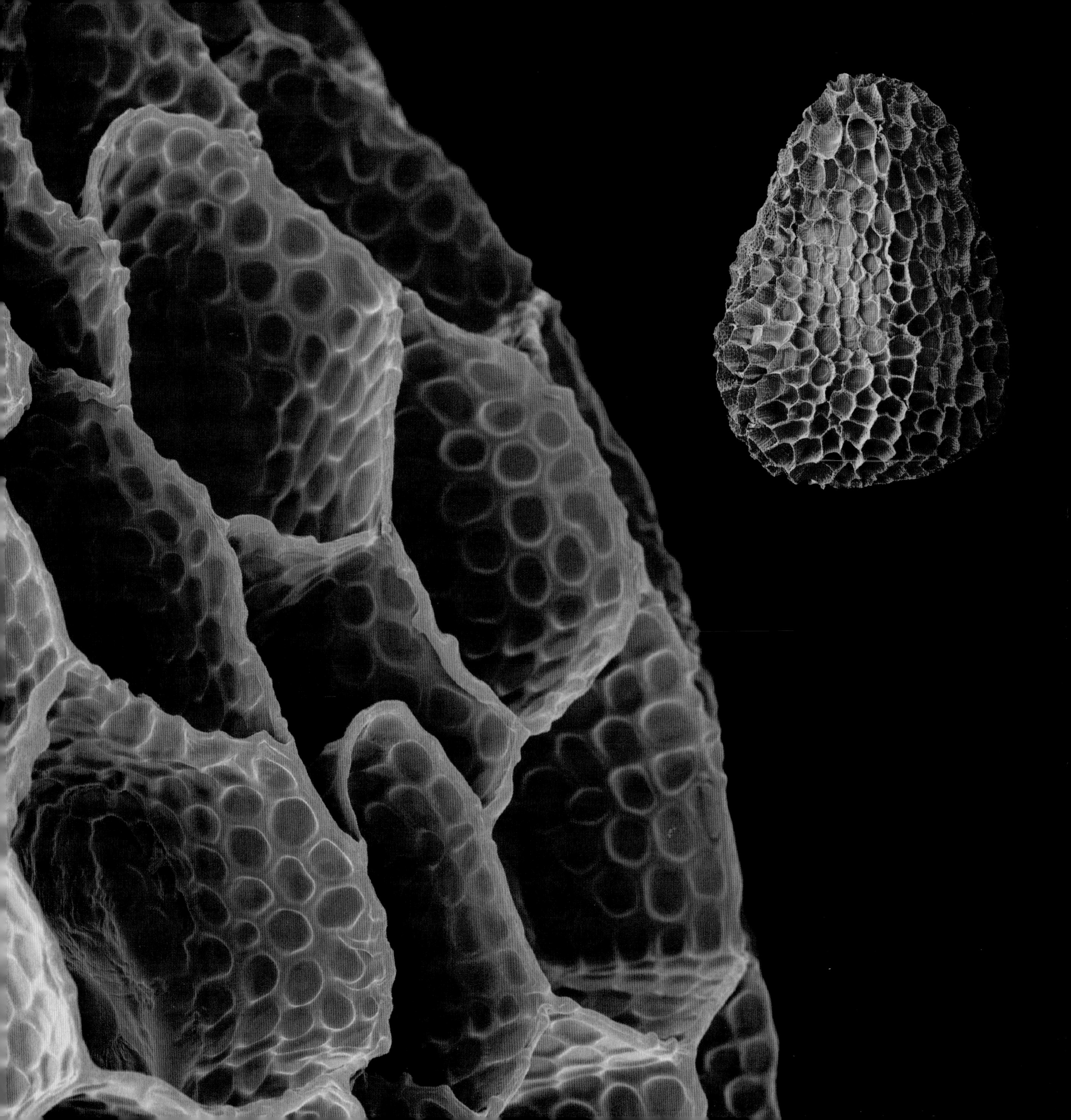

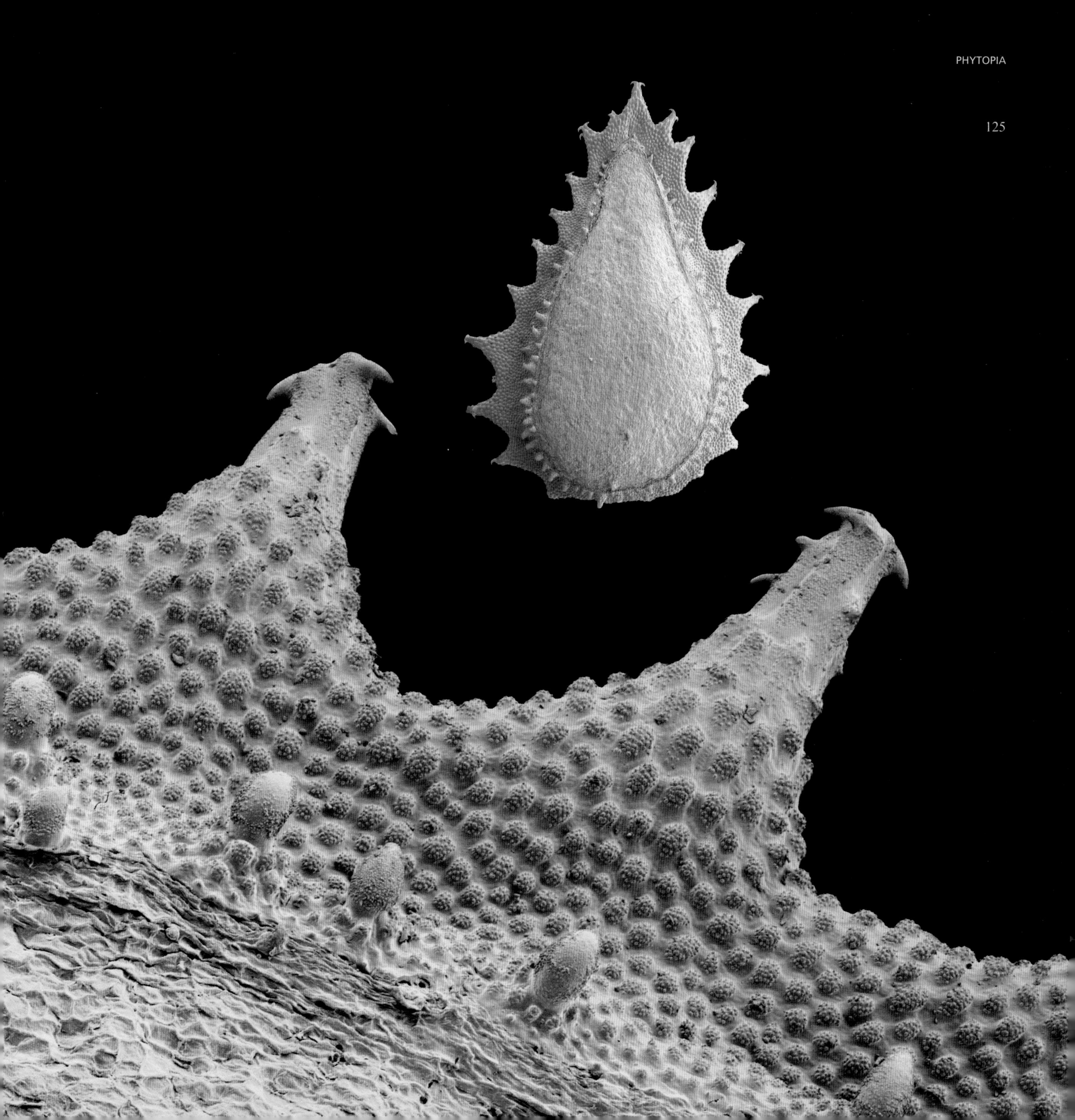

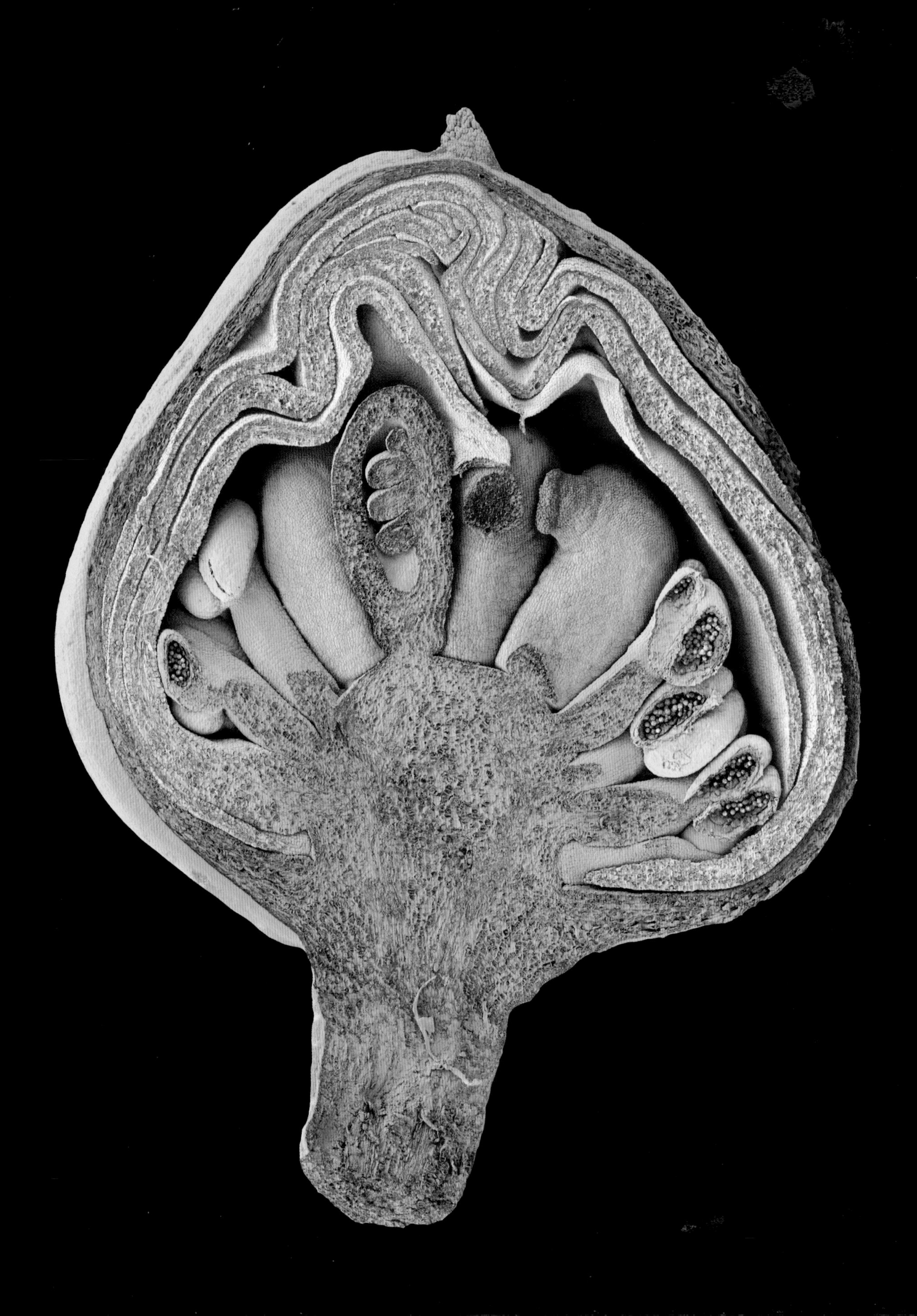

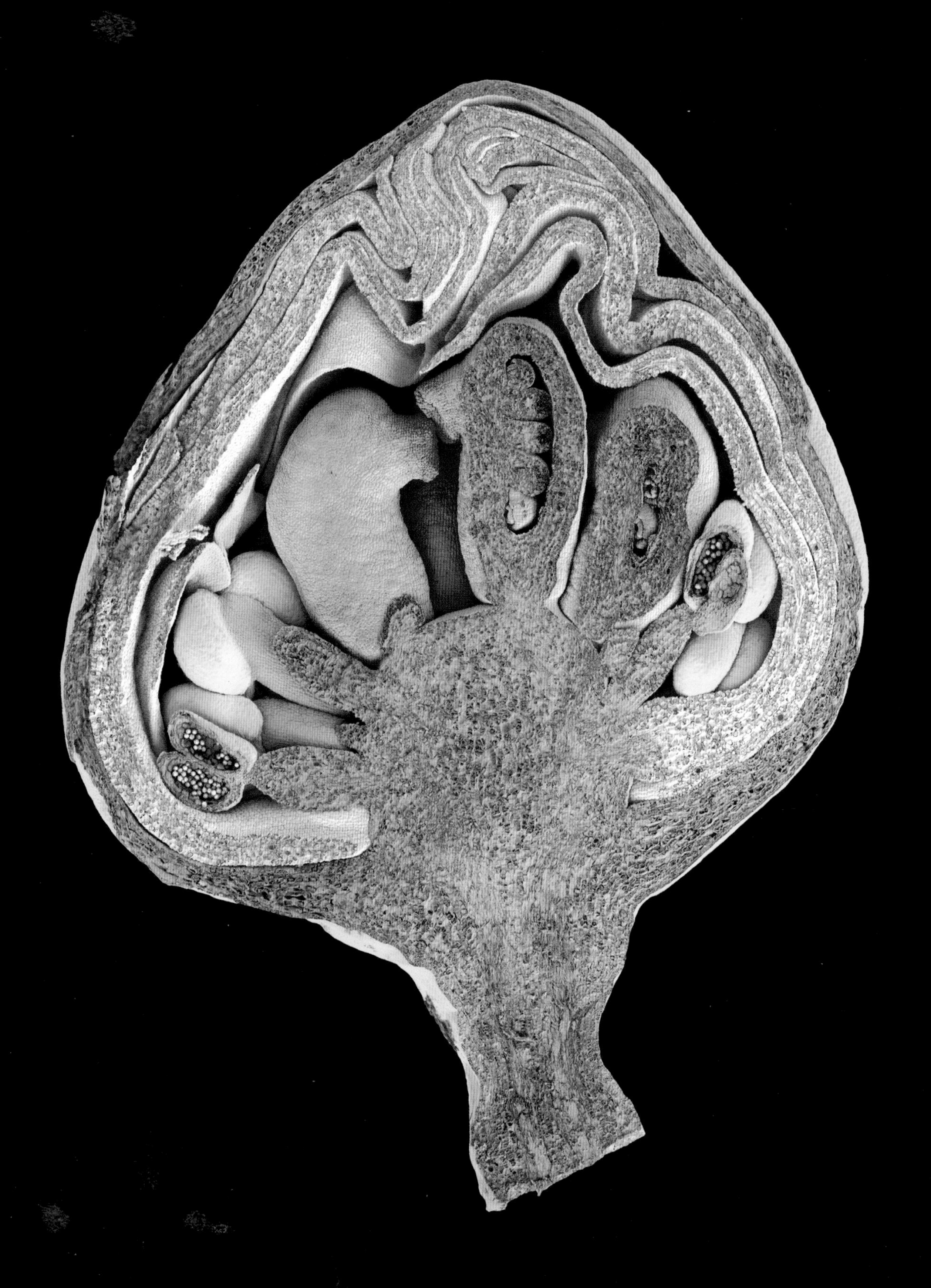

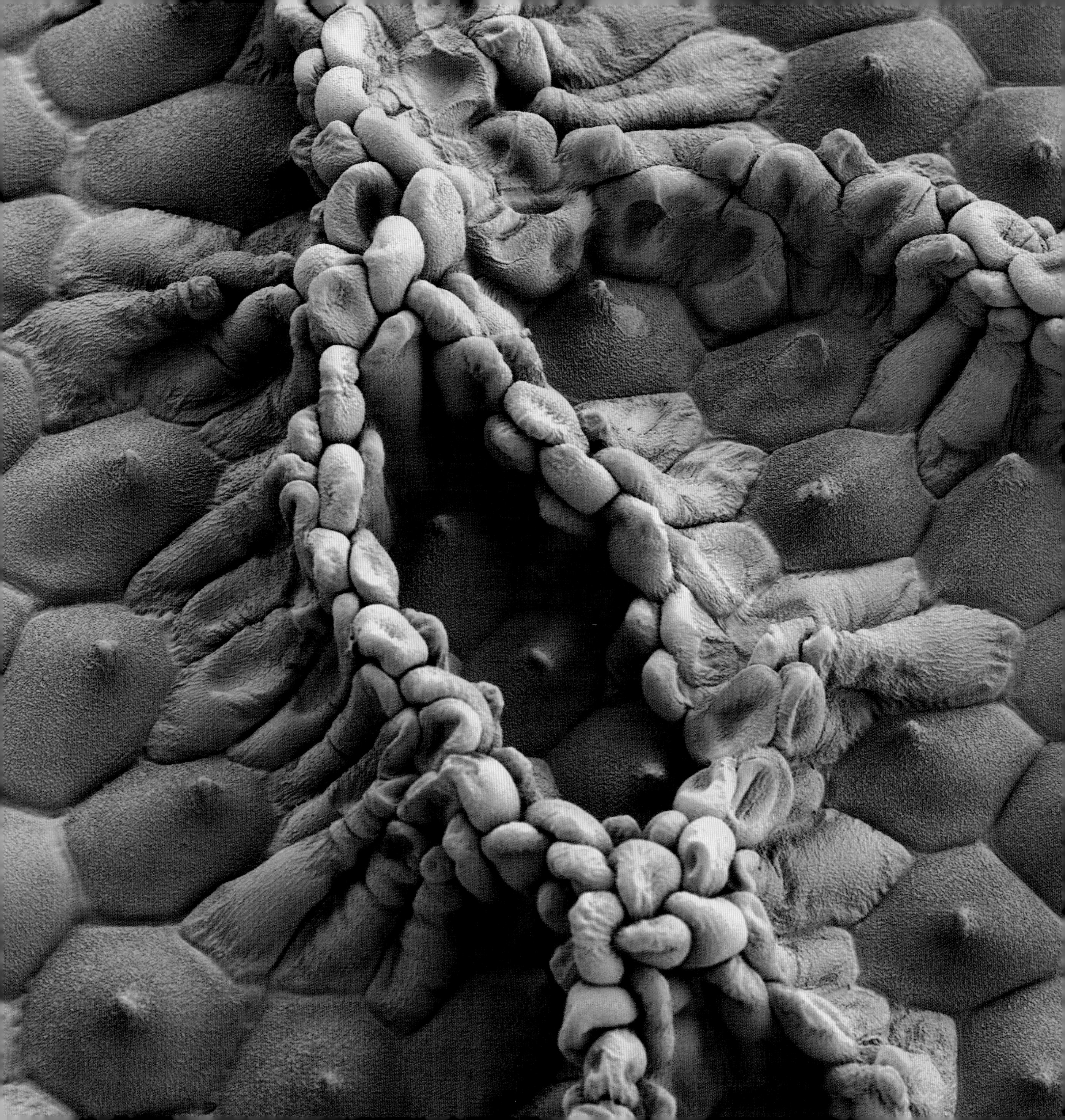

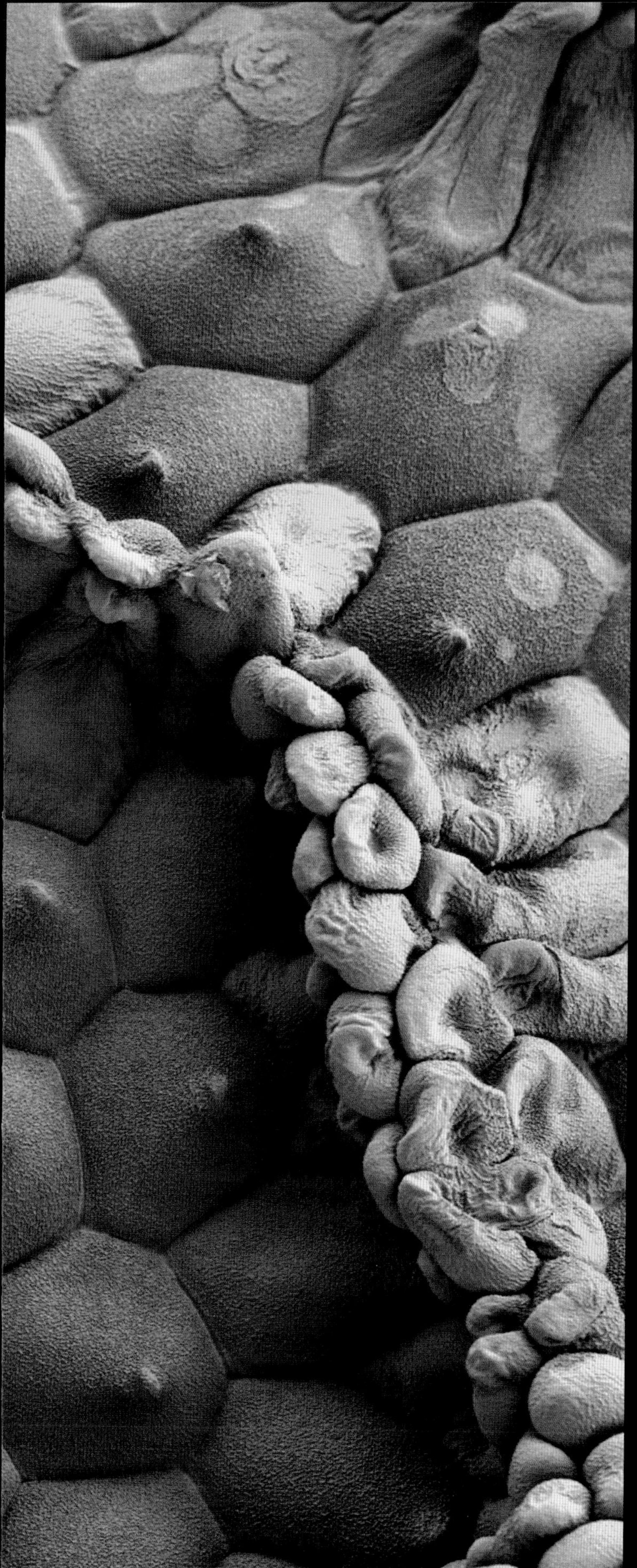

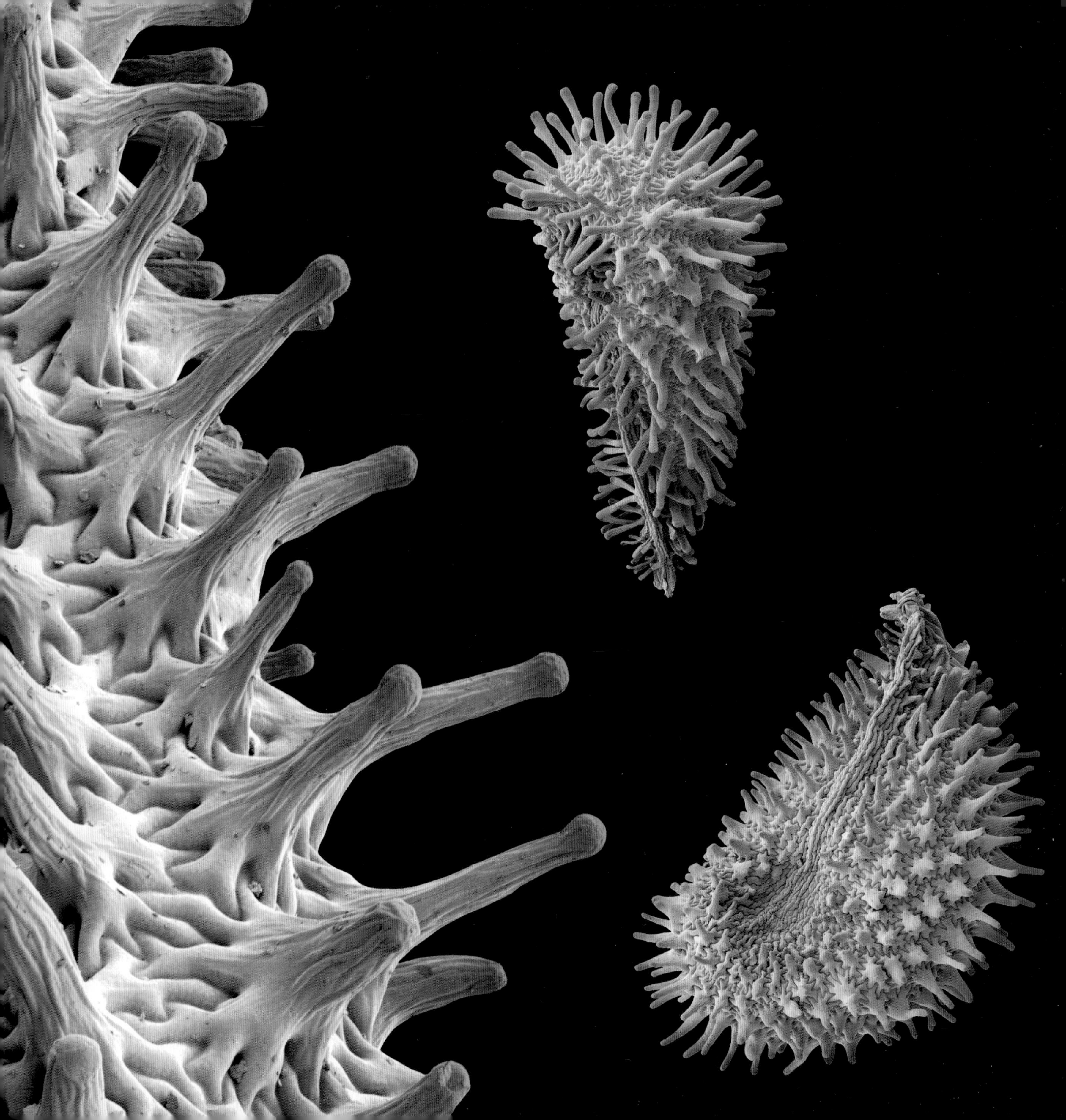

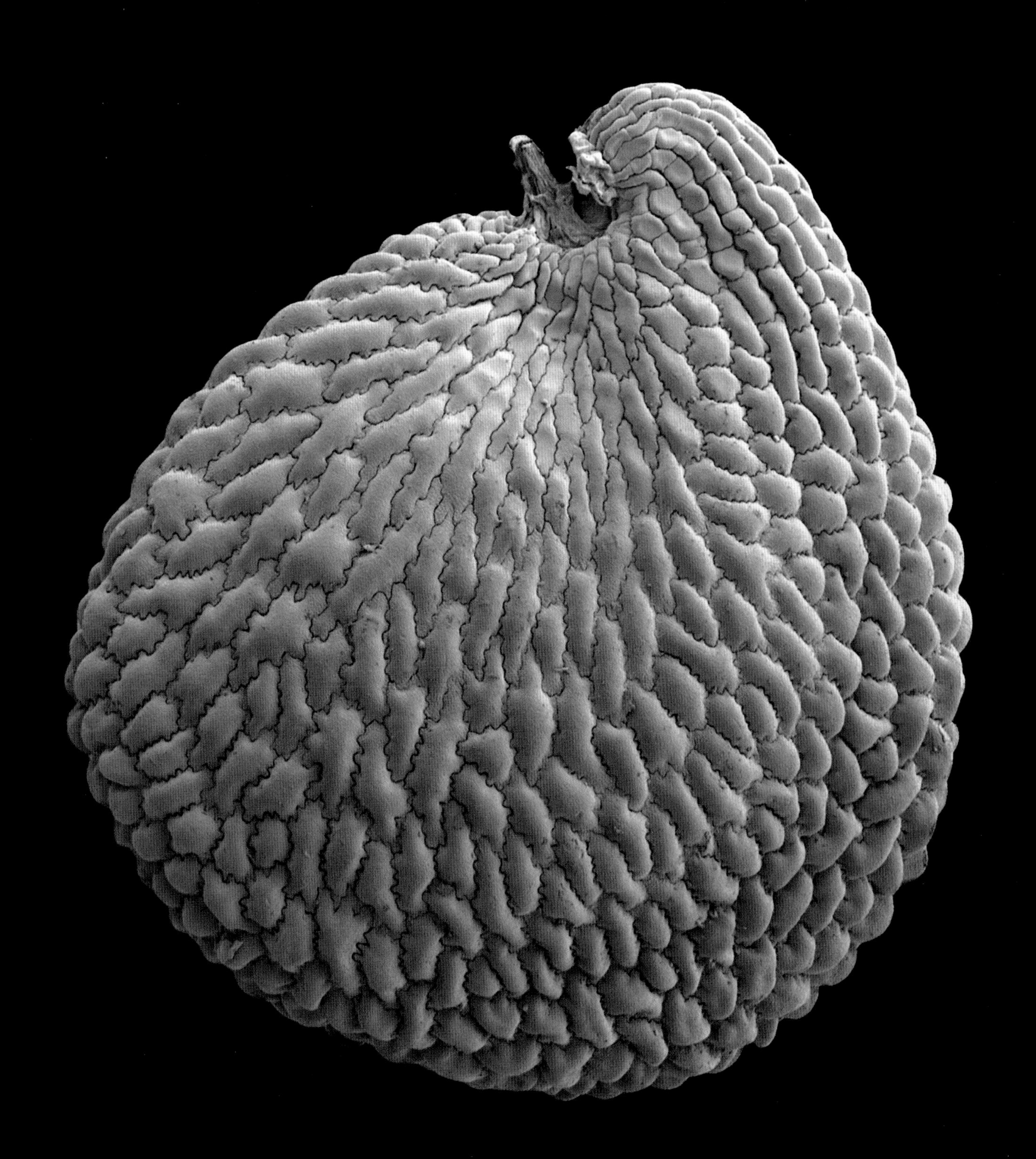

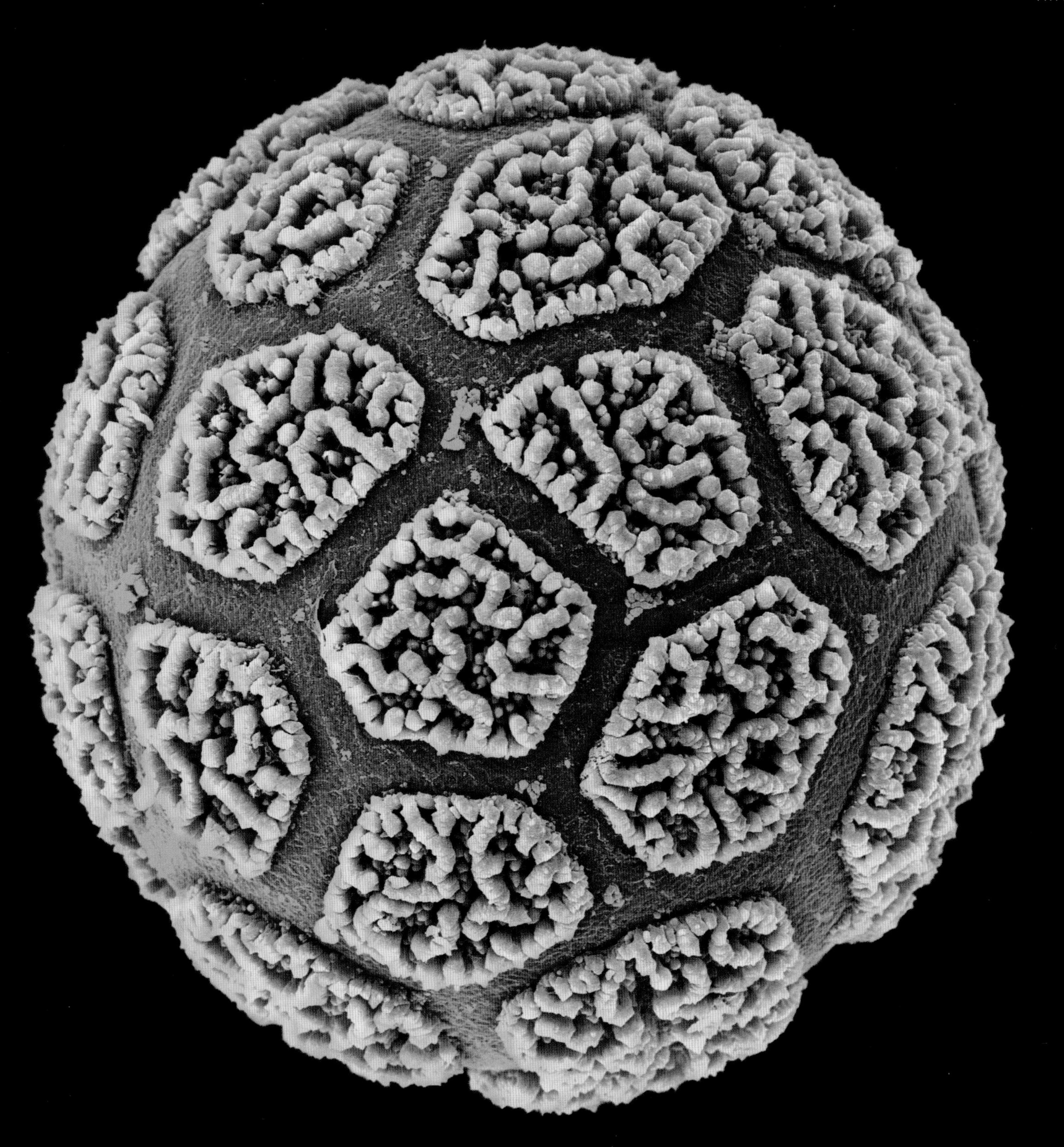

FOR PAGES WITH MULTIPLE IMAGES, DESCRIPTIONS READ CLOCKWISE FROM TOP LEFT.

Page 6: larkspur (*Delphinium peregrinum*, Ranunculaceae) – native to the Mediterranean; wind-dispersed seed covered by papery, lacerated lamellae; seed, diameter 1.5mm

Page 7: whorlflower (*Morina longifolia*, Morinaceae) – native to the Himalayas; pollen grain [SEM x4800]. Confederate rose (*Hibiscus mutabilis*, Malvaceae) – native to China and Japan but naturalised in the southern USA; as an adaptation to wind-dispersal, the seed bears a dorsal patch of spreading hairs that form a "parachute"; seed 2.6mm long (excluding hairs)

Pages 8/9: twining purslane (*Calandrinia eremaea,* Portulacaceae) – native to Australia and Tasmania; seeds, diameter 0.56 mm

Page 10: Jamaican poinsettia (*Euphorbia punicea*, Euphorbiaceae) – native to Jamaica; an exceptionally beautiful member of the spurge family; the yellow kidney-shaped structures are glands secreting nectar to attract insect[illegible] pollination

Page [illegible] horse-shoe vetch (*Hippocrepis unisiliquosa*, Legu[illegible]sae) – native to Eurasia and Africa; fruit; although the adaptive strategy behind the curiously shaped pods is difficult to interpret, its flat, lightweight construction may assist wind-dispersal. Moreover, the overlapping margins of the invaginat[illegible] the peripheral bristles may help attach the fruit to t[illegible] animals (epizoochory); diameter 18mm

Pag[illegible]3: *Bulbostylis hispidula* subsp. *pyriformis* (Cy[illegible]raceae) – native to East Africa; fruit without any apparent adaptations to a particular mode of dispersal; like many grasses (Poaceae) the plant may simply rely on herbivorous (plant-eating) animals accidentally ingesting the tiny fruits when browsing on the foliage, thereby facilitating dispersal; 1.3mm long. Sturt's desert pea (*Swainsona formosa*, Leguminosae) – native to Australia; famous for its striking blood-red flowers with a black centre, this species is one of Australia's most iconic wildflowers ('*formosa*' is Latin for 'beautiful'). The flowers are pollinated by birds

Page 14: pussy willow (*Salix caprea*, Salicaceae) – native to Eurasia; group of pollen grains [SEM x1500]

Page 1[illegible] eville orange (*Citrus aurantium*, Rutaceae) – native t[illegible]opical Asia; pollen grain with three apertures, 0.03mm long [SEM x2500]. Kaffir lime (*Citrus hystrix*, Rutaceae) – native to Indonesia; flowers with four white petals, numerous stamens and a prominent pistil (ovary green, style white, sti[illegible]a yellow); flowers, c. 14mm

Page 17: kaffi[illegible]me (*Citrus hystrix*, Rutaceae) – flower bud with petals and stamens partly removed to allow a view of the pistil; after fertilization the ovary produces a small knobbly gree[illegible]fruit, the kaffir lime; diameter 5.5mm

Pag[illegible]8/19: almond (*Prunus dulcis*, Rosaceae) – native to we[illegible] Asia; pollen grains germinating on agar medium [SEM x1000]

Page 20: gingerbush (*Pavonia spinifex*, Malvaceae) – spiny pollen grain typical of insect-pollinated flowers diameter 0.15mm [SEM x500]. Male fern (*Dryopteris filix-mas*, Dryopteridaceae) – single spore, 0.04mm long. *Osbeckia crinita* (Melastomataceae) – native to eastern Asia; seed, 0.65mm long

Page 21: corn poppy (*Papaver rhoeas*, Papaveraceae) – native to Eurasia and North Africa; pollen grain, 0.016mm [SEM x4000]. Manna ash (*Fraxinus ornus*, Oleaceae) – native to Eurasia; pollen grain [SEM x3500]. Meadow buttercup (*Ranunculus acris*, Ranunculaceae) – pollen grains, diameter 0.025mm [SEM x1800]

Page 22: male fern (*Dryopteris filix-mas*, Dryopteridaceae) – native to the temperate northern Hemisphere; underside of a fertile frond showing the brown sori (clusters of sporangia)

Page 23: False rose of Jericho (*Selaginella lepidophylla*, Selaginellaceae) – native to the Chihuahuan desert; a resurrection plant that can survive almost total desiccation; when exposed to humidity its tightly curled up leaves unfold. *Ampelopteris prolifera* (Thelypteridaceae) – native to the Old World Tropics; young sporophyte emerging from underside of liverwort-like gametophyte.

Page 24: Sichou oak (*Cyclobalanopsis sichourensis*, Fagaceae) – native to China; upon germination, the large acorn of this very rare species opens with a lid; diameter c. 4cm. Maidenhair tree (*Ginkgo biloba*, Ginkgoaceae) – two seedlings germinating from a single seed; c. 3cm. Sandbox tree (*Hura crepitans*, Euphorbiaceae) – native to South America and the Caribbean; seedling; the large discoid seed (c. 2cm) remains attached until its food reserves are used up. Acacia (*Acacia laeta*, Leguminosae) – native to Africa and the Middle East; germinating seed, c. 4cm long

Page 25: Yellow owl's clover (*Orthocarpus luteus,* Orobanchaceae) – native to North America; seed, 1.3mm long. Turk's-cap cactus (*Melocactus zehntneri,* Cactaceae) – native to Brazil; seed with part of the 'umbilical cord' (funicle) attached; 1.2mm long. *Garcinia arenicola* (Clusiaceae) – a Malagasy relative of the mangosteen; seedling, c. 10cm tall

Page 26: Japanese lily (*Lilium speciosum* var. *clivorum*, Liliaceae) – native to Japan; flower with large, pollen-laden anthers. *Nerine bowdenii* (Amaryllidaceae) – native to South Africa; pollen grain , 0.1mm long [SEM x1000]

Page 27: Rice's wattle (*Acacia riceana* hybrid, Leguminosae) – native to Tasmania; three pollen clusters (polyads), 0.035mm long [SEM x1500]

Page 28: angelim vermelho (*Dinizia excelsa*, Leguminosae) – native to Brazil and Guyana; pollen tetrad; diameter: 0.05mm. Comfrey (*Symphytum officinale*, Boraginaceae) – native to Europe; pollen grains with a series of apertures around the equator [SEM x2000]. Winter's bark tree (*Drimys winteri*, Winteraceae) – native to Chile and Argentina; pollen tetrad, diameter 0.04mm. Orchid tree (*Bauhinia* sp., Leguminosae) – pollen tetrad, diameter 0.08mm [SEM x800]

Page 29: lisianthus (*Eustoma grandiflorum*, Gentianaceae) – native to America and the Caribbean; monocolpate (with one elongated aperture); pollen grains on the surface of the anther, pollen grain, 0.016mm long [SEM x3500]

Page 30: motherwort (*Leonurus cardiaca*, Lamiaceae) – native to central Asia; two pollen grains showing characteristic 'ladder-like' rupturing of aperture membranes [SEM x3000]. Quince (*Cydonia oblonga,* Rosaceae) – cultivated since antiquity; view of pollen grain showing one of the three colpi (elongated apertures), the other two are out of view, 0.045mm long [SEM x2000]. Lenten rose (*Helleborus orientalis*, Ranunculaceae) – pollen grain with three colpi (elongated apertures), diameter: 0.034mm [SEM x2000]

Page 31: Judas tree (*Cercis siliquastrum*, Leguminosae) – native to southern Europe; pollen grains with three colpi (elongated apertures), length: 0.03mm [SEM x1500]

Page 32: *Persoonia mollis* (Proteaceae) – native to Australia; group of triporate (three round apertures) pollen grains [SEM x1000]. Mpumalanga Sagebush (*Hemizygia transvaalensis*, Lamiaceae) – native to South Africa; pollen grain with six colpi (elongated apertures) [SEM x1500]. Fruit-scented sage (*Salvia dorisiana*, Lamiaceae) – native to Honduras; pollen grain with three colpi (elongated apertures), 0.07mm long [SEM x1300]

Page 33: greater stitchwort (*Stellaria holostea*, Caryophyllaceae) – native to Europe; pollen grains, each bearing twelve round apertures, diameter 0.035mm [SEM x900]. Corncockle (*Agrostemma githago*, Caryophyllaceae) – native to Europe; multi-porate pollen grain; each pore is an aperture through which the developing pollen tube is potentially able to germinate, diameter 0.06mm [SEM x1500]

Page 34: Namaqualand in flower – after the spring rainfalls, the semi-desert landscape of South Africa's Namaqualand turns into a natural wonder unrivalled anywhere on Earth

Page 35: terracotta gazania (*Gazania krebsiana*, Asteraceae) – one of Namaqualand's most striking wildflowers

Page 36: black alder (*Alnus glutinosa*, Betulaceae) – native to Europe; catkins dispersing their pollen; note the old (last year's) female cones above. Hazelnut (*Corylus avellana*, Betulaceae) – native to Eurasia; typical of wind-dispersed pollen, the grains are smooth and free from sticky pollenkitt; [SEM x2000]. Rough meadow grass (*Poa trivialis*, Poaceae) – monoporate (with a single round aperture); pollen grain, diameter 0.055mm [SEM x1500]

Page 37: Monterey pine (*Pinus radiata*, Pinaceae) – native to California; two pollen grains, each bearing a pair of air sacs assisting wind-dispersal, 0.06mm wide [SEM x2000] Hazelnut (*Corylus avellana*, Betulaceae) – native to Eurasia; female flower exposing its red stigma-branches poised to catch pollen from the air; typical of wind-pollinated plants, the hazelnut bears tiny separate male and female flowers without showy perianths

Page 38: a Commissaris's long-tongued nectar bat (*Glossophaga commissarisi*, Phyllostomidae) pollinates the flower of *Markea neurantha*, a tropical member of the nightshade family (Solanaceae)

Page 39: flowers of the Australian red ash (*Alphitonia excelsa*, Rhamnaceae) visited by a bee

Page 40: *Rhododendron* cv. 'Naomi Glow' (Ericaceae) – close-up of flower; instead of being covered in sticky pollenkitt, the pollen grains are strung together by non-elastic, non-sticky, flexible 'viscin' fibres, which attach the grains to visiting insects

Page 41: lenten rose (*Helleborus orientalis*, Ranunculaceae) – native to Greece and Turkey; close-up of pollen group; note the sticky pollenkitt holding the pollen grains together [SEM x1000]

Page 42: *Orbea lutea* (Apocynaceae) – native to southern Africa; adapted to carrion fly pollination, flowers are fringed with hairs (mimicking the fur of a dead animal) and emit a putrid smell

Page 43: bee orchid (*Ophrys apifera*, Orchidaceae) – native to Europe and North Africa; the flowers mimic the appearance of a female bee attracting male bees, which pollinate the flowers while trying to copulate with them. Indian pokeweed (*Phytolacca acinosa*, Phytolaccaceae) – native to east Asia; flower, diameter 7.5mm

Page 44: a European honey bee (*Apis mellifera*) visits a cornflower (*Centaurea cyanus*, Asteraceae). *Rudbeckia* hirta

'Prairie Sun' (Asteraceae) – capitulum (flower head) as seen under normal conditions; the same capitulum as seen under ultaviolet light, revealing the typical 'bulls-eye' pattern of nectar guides as bees see it

Page 45: Pollen of common mallow (*Malva sylvestris*, Malvaceae) on the rear leg of a bumblebee (*Bombus terrestris*) [SEM x100]

Page 46: horse chestnut (*Aesculus hippocastanum*, Sapindaceae) – native to south-east Europe; flower and pollen grains [SEM x3000]; zygomorphic flowers with coloured spots on the petals that serve as nectar guides which are typical for the bee-pollination syndrome

Page 47: red campion (*Silene dioica*, Caryophyllaceae) – native to Europe; flower and pollen grains [SEM x2000]; a flat, plate-like corolla (landing platform), red in colour and a long, narrow flower tube with deeply hidden nectar, are typical floral features of the butterfly-pollination syndrome

Page 48: Monarch butterfly on a lantana flower (*Lantana camara*)

Page 49: Pollination of the Malagasy comet orchid (*Angraecum sesquipedale*) through the giant hawkmoth *Xanthopan morganii praedicta*, the only insect with a tongue long enough to reach the nectar at the end of the huge spur (30-35 cm long). Yellow tailflower (*Anthocercis ilicifolia*, Solanaceae) – native to Western Australia; not much is known about the identity of the pollinators of this flower but the subtle scent indicates moth pollination

Page 50: *Impatiens tinctoria* (Balsaminaceae) – native to Africa; with their dark red spotted throats, long nectar spur and pleasant nocturnal scent, the flowers of this tropical balsam are clearly adapted to moth-pollination

Page 51: snake gourd (*Trichosanthes cucumerina*, Cucurbitaceae) – native to Asia; strictly adapted to moth-pollination, the strongly scented, lace-like white flowers of this tropical and subtropical climber from the pumpkin family unfurl for only one night; the plant is grown for its very long, snake-like fruit, which is used as a vegetable in Asia

Page 52: devil's cotton (*Abroma augusta*, Malvaceae) – native to Asia and Australia; the purple-brown, lantern-like flowers are pollinated by tiny freeloader flies (family Milichiidae) whose larvae scavenge on organic matter in the nests of ants and birds

Page 53: *Huernia hislopii* (Apocynaceae) – native to Africa; a typical carrion-flower, mimicking the looks and smell of the rotting flesh in which carrion flies normally lay their eggs. As the deceived flies deposit their eggs in the throat of the flowers (note the white lumps of eggs), they unwittingly act as pollinators. Californian allspice (*Calycanthus occidentalis*, Calycanthaceae) – endemic to California; the large robust flowers of this shrub are pollinated by beetles

Page 54: Brown-throated sunbird (*Anthreptes Malaconsis*)

Page 55: Elim heath (*Erica regia*, Ericaceae) – native to South Africa's unique fynbos vegetation; pendant tubular red flowers are telltale signs of the bird-pollination syndrome. The plant's strong branches are able to support sunbirds that visit to feed on the nectar of the flowers. Waratah (*Telopea speciosissima*, Proteaceae) – endemic to New South Wales in Australia; the robust texture of the inflorescences and the striking red colour are clear indicators of the bird-pollination syndrome. In its native New South Wales, honeyeaters (birds of the family Meliphagidae) are the main pollinators

Page 56: sausage tree (*Kigelia africana*, Bignoniaceae) – native to tropical Africa – the fragrant, maroon-red, nectar-rich flowers of this African tree dangle down on long, rope-like stalks where they can easily be reached by bats. Although bats are the main pollinators, insects and sunbirds also visit the flowers

Page 57: baobab (*Adansonia digitata*, Malvaceae) – native to Africa; the 10-20cm large white, sweetly scented nocturnal flowers dangle down on long stalks where bats can easily reach them to feed on the copious nectar they contain. The powder-puff-like arrangement of the stamens is typical of bat-pollinated flowers and ensures that the furry visitors are well dusted with pollen

Page 58: honey possum (*Tarsipes rostratus*, Tarsipedidae) – this tiny Australian marsupial is a highly adapted pollinator living entirely off the nectar and pollen it collects from flowers, especially the long narrow flowers of the protea family (Proteaceae). Its narrow projecting snout with very few teeth or none at all, and a long tongue with a brush-like tip are clear adaptations to its specialised lifestyle

Page 60: *Fragaria* x *ananassa* (Rosaceae) – garden strawberry; only known in cultivation; young fruit– fruit (glandetum); in strawberry flowers many separate carpels are arranged on a convex floral axis. As the flower turns into a fruit, the axis grows into the fleshy, edible part of the fruit. The carpels themselves turn into tiny, brown nutlets which are sunken into the fleshy receptacle. The persistent styles of the individual carpels are responsible for the bristly texture of strawberries; diameter 1.2cm

Page 61: *Citrus margarita* (Rutaceae) – kumquat; cultivated for centuries, probably originating in southern China; cross section through fruit; the edible part of citrus fruits consists of little 'juice sacs' that emerge from the inner surface of the ovary wall; diameter 2.1cm

Page 62: natal sundew (*Drosera natalensis*, Droseraceae) – native to southern Africa and Madagascar; seed, 0.8mm long. Sea campion (*Silene maritima,* Caryophyllaceae) – native to Europe; seed, 1.3mm long. Small-flowered catchfly (*Silene gallica*, Caryophyllaceae) – native to Eurasia and North Africa; seed, 1.5mm long. *Crassula pellucida* (Crassulaceae) – native to South Africa; seed, 0.8mm long

Page 63: prickly starwort (*Stellaria pungens*, Caryophyllaceae) – native to Australia; seed, 1.5mm long. Yellow paintbrush (*Castilleja flava*, Orobanchaceae) – native to North America; seed, 1.5mm long

Page 64: Lake Logue wattle (*Acacia vittata,* Leguminosae) – native to south-western Australia; fruit and seeds; like many wattles (Australian acacias), the seeds of the Lake Logue wattle are equipped with a 'bait' (elaisome) to attract ants for dispersal; fruit, 21mm long, seed 3.8mm long. Coastal wattle (*Acacia cyclops,* Leguminosae) – native to south-western Australia; seed surrounded by a brightly orange-coloured aril to attract birds for dispersal; seed, 9mm long

Page 65: scarlet pimpernel (*Anagallis arvensis*, Myrsinaceae) – native to Europe; the capsule of the scarlet pimpernel opens with a lid to let the seeds tumble out. The stiff persistent style at the apex may assist in the removal of the lid as the capsule is touched by a passing animal or brushes against other plants when moved by wind; fruit, diameter 4mm

Page 66: greater sea-spurrey (*Spergularia media,* Caryophyllaceae) – native to Eurasia and North Africa; wind-dispersed seed with peripheral wing assisting wind-dispersal; diameter 1.5mm. *Galinsoga brachystephana* (Asteraceae) – native to Central and South America; in the tiny shuttlecock-like fruits of this species the rays of the modified calyx perform as tiny feathery wings; 2.5mm long. American bugbane (*Cimicifuga americana,* Ranunculaceae) – native to eastern North America; the bizarre lobes of the seed are most likely an adaptation to wind-dispersal; 4.3mm long. Leeubekkie (Afrikaans) (*Nemesia versicolor*, Plantaginaceae) – native to South Africa; seed with peripheral wing assisting wind-dispersal.

Page 67: firebush (*Hymenodictyon floribundum,* Rubiaceae) – native to Africa; wafer-thin, wind-dispersed seed with a peripheral wing; 8.2mm long

Page 68: rosebay willowherb (*Epilobium angustifolium,* Onagraceae) – native to the Northern Hemisphere; seed with a tuft of hairs assisting wind-dispersal; 0.95mm long (without hairs)

Page 69: crown flower (*Artedia squamata*, Apiaceae) – endemic to Cyprus and the eastern Mediterranean; flat, wind-dispersed fruit with a series of peripheral wings; 1cm long. Crowned lamb's lettuce (*Valerianella coronata*, Valerianaceae) – native to the Mediterranean and south-west and central Asia; fruit with enlarged parachute-like calyx with tips drawn out to form hooked spines that assist both wind- and animal-dispersal; diameter 5.2mm. Princess tree (*Paulownia tomentosa*, Paulowniaceae) – native to China; seed equipped with a lobed peripheral wing assisting wind-dispersal; 4.4mm long. *Scabiosa crenata* (Dipsacaceae) – native to the Mediterranean; the fruit of this species pursues a double dispersal strategy: the papery collar facilitates wind-dispersal whereas the rough calyx awns are poised to hook onto the fur of passing animals; diameter 7.2mm

Page 70: Examples of tiny, dust-like seeds:
Top left: *Blossfeldia liliputana* (Cactaceae) – native to Argentina and Bolivia; seed with elaiosome to assist dispersal by ants; 0.65mm long. Fully grown at just 12mm this species is the smallest cactus in the world. Top right: cistus-flowered sundew (*Drosera cistiflora,* Droseraceae) – native to South Africa; seed, 0.5mm long. Centre: pink sundew (*Drosera capillaris*, Droseraceae) – native to the eastern United States; seeds, 0.6mm long. Centre left: rock sea-spurrey (*Spergularia rupicola,* Caryophyllaceae) – native to Europe; seed, 0.6mm long. Bottom right: London pride (*Saxifraga umbrosa*, Saxifragaceae) – endemic to the Pyrenees; seed, 0.6mm long. Bottom left: piggyback plant (*Tolmiea menziesii,* Saxifragaceae) – native to Oregon; seed, diameter 0.6mm.

Page 71: Bell heather (*Erica cinerea*, Ericaceae) – native to Europe and North Africa; seed, 0.7mm long. Tiger-like stanhopea (*Stanhopea tigrina*, Orchidaceae) – native to tropical America; tiny wind-dispersed seeds with a loose bag-like seed coat; 0.66mm long. Broomrape (*Orobanche* sp., Orobanchaceae) – collected in Greece; seeds, 0.35-0.4mm long

Pages 72/73: hoary sunray (*Leucochrysum molle*, Asteraceae) – native to Australia; detail of one pappus ray with a pollen grain accidentally stuck to it; pollen grain, diameter 0.025mm

Page 74: Examples of seeds with extreme honeycomb patterns: *Loasa chilensis* (Loasaceae) – native to Chile; seed, 1.9mm long. Purple owl's clover (*Castilleja exserta* subsp. *latifolia,* Orobanchaceae) – native to California; seed, 1.9mm long. *Lamourouxia viscosa* (Orobanchaceae) – native to Mexico; seed, 1.2mm long

Page 75: *Loasa chilensis* (Loasaceae) – detail of seed coat [SEM x150]

Page 76: lesser snapdragon (*Antirrhinum orontium,* Plantaginaceae) – native to Europe; fruit opening with irregular ruptures at the top. The seeds are dispersed like salt from a salt shaker as the capsules sway in the wind or are moved by passing animals. The stiff, spike-like remnant of the style may assist with the latter; fruit, 7mm long. Red Campion (*Silene dioica*, Caryophyllaceae) – native to Europe; fruit and seed dispersal occurs when the capsules expel the seeds as they sway in the wind; seed, 1.2mm long

Page 77: seeds of Red Campion (*Silene dioica*, Caryophyllaceae). Corn poppy (*Papaver rhoeas*, Papaveraceae) – native to Eurasia and North Africa; capsule; the seeds are flung out as the capsule sways in the wind on its long flexible stalk; diameter 6.5mm. Lesser snapdragon (*Antirrhinum orontium,* Plantaginaceae) – native to Europe; seeds, 1.1mm long

Page 78: sea mango (*Cerbera manghas*, Apocynaceae) – native from the Seychelles to the Pacific; drift fruit commonly found as flotsam on the beaches of the Indian and Pacific Oceans. A massive fibrous, corky mesocarp affords excellent, long-lasting buoyancy in seawater; fruit, 9cm long. Nipa palm (*Nypa fruticans,* Arecaceae) – native from southern Asia to northern Australia; longitudinal section through the single-seeded coconut-like fruit; the seed inside germinates before it is dispersed; the emerging pointed shoot assists in the detachment from the parent plant. Between the hard outer, seawater-resistant epicarp, and the bony endocarp lies a fibrous-spongy mesocarp that acts as a buoyancy device; 11.5cm long

Page 79[illegible]ooking-glass mangrove (*Heritiera littoralis*, Malvac[illegible]) – native to the Old World tropics; the seawater-pr[illegible]-like fruit contains a single round seed surrounded [illegible]rge air space. The prominent keel on the back acts li[illegible]il; fruit, up to 10cm long. Yellow floatingheart (*Nymphoides peltata*, Menyanthaceae) – native to Eurasia; water-dispersed seed and close-up of marginal bristles. Although heavier than water, the flat shape, water-repellent surface and fringe of stiff bristles allow the seeds to use the surface tension of water to avoid sinking; seed, 5mm long

Page 80: collection of drift 'seeds' containing fruits of the looking-glass mangrove (*Heritiera littoralis*, Apocynaceae) and various 'sea beans' such as the legendary Mary's or crucifixion bean (*Merremia discoidesperma*, Convolvulaceae), characteristically marked by a cross, and many legume seeds, including the sea heart (*Entada gigas*), hamburger bean (*Mucuna* spp.), grey nickernut (*Caesalpinia bonduc*) and sea purses (*Dioclea* spp.)

Page 81: yellow nickernut (*Caesalpinia major*, Leguminosae) – found in the tropics worldwide; seed, 2.5cm long. Hamburger bean (*Mucuna urens*, Leguminosae) – native to Central and Sou[illegible]erica, diameter 2.5cm. Sea heart (*Entad*[illegible]*gas*, [illegible]nosae) – native to tropical America and Africa; this very [illegible]on 'sea bean' comes from the gigantic pods of a tropical liana up to 1.8m long

Page 82: Seychelles nut (*Lodoicea maldivica*, Arecaceae) – native to the Seychelles Islands; the single-seeded fruits of this palm tree take 7-10 years to mature and contain the world's largest seed. Nato mangrove (*Mora megistosperma*, Leguminosae) – native to tropical America; individual seed and opened fruit with two seeds; the nato mangrove seed is up to 18cm long and weighs nearly a kilogram; it is the largest seed of all dicotyledonous flowering plants

Page 83: Chinese winter hazel (*Corylopsis sinensis* var. *calvescens,* Hamamelidaceae) – native to China; the capsular fruit opens slowly and as the extremely hard endocarp dries out it changes its shape, gripping the seeds like a vice. Eventually, the two hard, smooth, spindle-shaped seeds are forcibly expelled; fruit, diameter 7mm. Himalayan balsam (*Impatiens glandulifera*, Balsaminaceae) – native to the Himalayas; exploded fruit; when ripe the slightest touch can trigger the fruits to explode and hurl their small black seeds up to 5m away

Page 84: a Gambian epauletted fruit bat (*Epomophoros gambianus*, Pteropodidae) takes off with a fig. Along with birds and monkeys, fruit bats are the most important seed-dispersing animals in tropical rainforests

Page 85: villous fig (*Ficus villosus,* Moraceae) – native to tropical Asia; longitudinal section of fruit, and fruits on the tree. The c. 750 members of the genus *Ficus* (figs) bear their tiny flowers inside a peculiar inflorescence called a *syconium*, which after pollination matures into the fruit that we commonly refer to as a 'fig'. Morphologically a syconium can be compared to a sunflower head curving up its margin to first form a bowl and then an urn, leaving a small opening (ostiole) at the top. The entrance of the fig cavity is closed by numerous tightly packed bracts which, at the time of pollination, give way to a narrow passage through which the fig's pollinators, tiny fig wasps of the family Agaonidae, can enter the flower-lined cavity. In *Ficus villosa* and other species, the cavity of the syconium is filled with a mucilaginous fluid prior to pollination; fruit, diameter 12mm

Page 86: American stickseed (*Hackelia deflexa* var. *americana*, Boraginaceae); native to North America; single-seeded nutlet covered with hooked spines which very effectively attach this diaspore to plumage, fur and clothes. As in many members of the borage family, the ovary of *Hackelia* is deeply four-lobed and splits at maturity into four single-seeded nutlets; nutlet, 3.5mm long

Page 87: stickywilly (*Galium aparine,* Rubiaceae) – native to Eurasia and America; the fruit of the stickywilly is formed by two united carpels that break apart into two separate nutlets when ripe. The small branch has two very young developing fruits with their ovaries lobed but still entire and a flower bud consisting of the tiny ovary crowned by the closed perianth. Densely covered in tiny hooks, the nutlets of the stickywilly are tenacious hitchhikers; mature nutlets 5mm long

Page 88: Pima rhatany (*Krameria erecta,* Krameriaceae) – native to the southern USA and northern Mexico; the barbed spines covering the single-seeded fruit of this small shrub are a clear adaptation to ensure dispersal by attachment to the fur of passing animals (epizoochory); fruit, 8mm long (without spines)

Page 89: wild carrot (*Daucus carota,* Apiaceae) – native to Europe and south-western Asia; the small fruits of the wild carrot are covered in long spines with recurved hooks at their tip, a clear adaptation to facilitate dispersal by attachment to furry or feathery animals (epizoochory); 5.5mm long

Page 90: common agrimony, cockleburr (*Agrimonia eupatoria,* Rosaceae) – native to the Old World; the hooked spines covering the fruit are a very efficient dispersal aid, readily catching onto animal fur or clothes; 7.5mm long

Page 91: Coastal sandbur (*Cenchrus spinifex*, Poaceae) – native to America; the painfully spiny fruits of this abundant grass can cause problems for grazing lifestock; 9.5mm long (including spines). Tarara amarilla (*Centrolobium microchaete*, Leguminosae) – native to South America; with its spine-covered, seed-bearing part enables this winged nut (samara) to defend itself against predators but also pursue a double dispersal strategy (both wind and animal; c. 20cm long. Creeping carrot (*Trachymene ceratocarpa*, Araliaceae; formerly Apiaceae) – native to Australia; the peculiar fruitlets of this species bear two apical wings that assist wind-dispersal, as well as two dorsal rows of spines to assist animal dispersal (epizoochory); 4.5mm long. Burclover (*Medicago polymorpha,* Leguminosae) – native to Eurasia and North Africa; typical of the genus *Medicago*, the fruit is coiled into a spiral of 4-6 turns. With its spherical shape and hooked spines it is well adapted to attach itself to furry or feathery animals; diameter 9.5 mm (including spines)

Page 92: Golden devil's claw (*Proboscidea altheifolia,* Martyniaceae) – native to the southern USA and Mexico; after the soft green husk has dropped off the exposed woody core of the fruit splits down the middle, so that its beak produces two sharply pointed, recurved spurs, waiting for an animal to become entangled and carry the fruit away; fruit, 12cm long. *Uncarina* sp. (Pedaliaceae) – collected in Madagascar; probably the most tenacious of all fruits. Once caught in the extremely sharp, curved hooks that crown the tip of the fruit's long radiating spines it is almost impossible to escape without injury; fruit, diameter 8cm. Threecornerjack (*Emex australis*, Polygonaceae) – native to southern Africa; formed by the hardened calyx of the flower and arranged like a medieval caltrop, the ferocious spines of this fruit are poised to bury themselves into the skin of animals, a rather cruel method of animal dispersal; 8mm long.

Page 93: devils' claw, grapple plant (*Harpagophytum procumbens*, Pedaliaceae) – native to southern Africa and Madagascar; the large woody grapples of this devil's claw are adapted to cling to the feet and fur of animals, who may suffer terrible wounds; fruit, 9cm long. Devil's thorn (*Tribulus terrestris,* Zygophyllaceae) – native to the Old World; the fruit of the puncture vine splits into five single-seeded nutlets, one of which is shown here. Each nutlet is armed with two large and several smaller spines which are arranged like a medieval caltrop, poised to penetrate the skin of an animal or the soles of shoes; 6mm long

Page 94: many plants in temperate deciduous forests in Europe and North America and especially in the dry shrublands of Australia and South Africa have seeds that bear an edible fatty nodule (elaiosome) to attract ants. Here, western harvester ants (*Pogonomyrmex occidentalis*), a common species in the deserts and grasslands of western North America and northern Mexico, are busy carrying the seed of *Cnidoscolus* sp. (Euphorbiaceae) back to their nest where they remove the elaiosome to feed it to their larvae.

Page 95: seeds of the quinine bush (*Petalostigma pubescens*, Picrodendraceae) – native to Malesia and Australia; the yellowish appendages (elaiosomes) attract ants which take the seeds to their nest where they are safe from seed-eating rodents and seasonal fires; seeds, 1.2cm long

Page 96: seeds that bear fatty nodules (elaisomes) to attract ants for dispersal are found in over 80 plant families. The following are some examples: Top left: button creeper (*Tersonia cyathiflora*, Gyrostemonaceae) – native to Western Australia; seed, 2.7mm long. Top right: Aztec cactus (*Aztekum ritteri*, Cactaceae) – native to Mexico, seed, 0.8mm long. Centre right: petty spurge (*Euphorbia peplus,* Euphorbiaceae) – native to Eurasia; seed, 1.6mm long. Bottom right: Lake Logue wattle (*Acacia vittata,* Leguminosae) – native to south-western Australia; seed, 3.8mm long. Bottom left: sun spurge (*Euphorbia helioscopia,* Euphorbiaceae) – seed, 2.3mm long

Page 97: Top left: spurge (*Euphorbia* sp., Euphorbiaceae) – collected in Lebanon; seed, 3mm long. Top middle: poppywort (*Stylophorum diphyllum*, Papaveraceae) – native to the Eastern United States; seed, 2.2mm long. Top right: petty spurge (*Euphorbia peplus,* Euphorbiaceae) – native to Eurasia; seed, 1.6mm long. Bottom right: sand milkwort (*Polygala arenaria*, Polygalaceae) – native to tropical Africa; seed, 2.2mm long. Bottom left: *Blossfeldia liliputana* (Cactaceae) – native to Argentina and Bolivia; seed, 0.65mm long

Page 98: Japanese wineberry (*Rubus phoenicolasius*, Rosaceae) – native to northern China, Korea and Japan; a relative of the raspberry (*Rubus idaeus*) and the blackberry

(*Rubus fruticosus*) that also produces sweet and juicy edible fruits. Curiously, the whole plant, including the calyx surrounding the fruit is covered in sticky, glandular hairs; diameter c. 1cm
Page 99: garden strawberry (*Fragaria* x *ananassa*, Rosaceae) – only known in cultivation; famed for its delicate flavour and high vitamin content, the garden strawberry is one of the most popular fruits with an annual world production of more than 2.5 million tons; 3cm long
Page 100: examples of sweet and juicy fruits: Top left: tree tomato (*Solanum betaceum*, Solanaceae); diameter 4cm. Top middle: kiwi (*Actinidia deliciosa*, Actinidiaceae); diameter 4cm. Top right: papaya (*Carica papaya*, Caricaceae); 12cm long. Centre: pomelo (*Citrus maxima*, Rutaceae), diameter 15cm. Centre right: purple passionfruit (*Passiflora edulis* forma *edulis*, Passifloraceae); diameter 4cm. Centre left: *Galia muskmelon* (*Cucumis melo* subsp. *melo* var. *cantalupensis* 'Galia', Cucurbitaceae) – diameter c. 16cm. Bottom left: pomegranate (*Punica granatum*, Lythraceae); diameter 11 cm. Bottom right: dragon fruit (*Hylocereus undatus*, Cactaceae), c. 16cm long
Page 101: Top left: *Galia muskmelon* (*Cucumis melo* subsp. *melo* var. *cantalupensis* 'Galia', Cucurbitaceae); diameter c. 16cm. Top middle: fig (*Ficus carica*, Moraceae); diameter 4cm. Top right: peach (*Prunus persica* var. *persica*, Rosaceae); diameter 6cm. Centre right: Chinese pear (*Pyrus pyrifolia*); diameter 8cm. Centre left: mango (*Mangifera indica*, Anacardiaceae); 10cm long. Bottom right and middle: durian (*Durio zibethinus*, Malvaceae); 25cm long. Bottom left: lychee (*Litchi chinensis* ssp. *chinensis*, Sapindaceae); 3cm long
Pages 102/103: Chinese dogwood (*Cornus kousa* subsp. *chinensis*, Cornaceae) – native to central and northern China; the edible fleshy fruit develops from a spherical cluster of flowers and consists of a bright red head of coalescent drupes; diameter c. 2cm. The microscopic detail of an immature fruit shows an individual flower, its hairy calyx surrounding the ovary (the stamens have already dropped); style, 1mm long
Pages 104/105: peach (*Prunus persica* var. *persica*, Rosaceae) – originally from China; fruit (drupe) whole and cut in half. Microscopic detail of fruit surface; the downy texture of peach skin is due to thousands of trichomes (hairs), the majority of which are very short, stomata (breathing pores) marked red; pictured area, 0.7mm wide
Pages 106/107: Black mulberry (*Morus nigra*, Moraceae) – cultivated since antiquity, originally probably from China; although similar in appearance to a blackberry or raspberry, the black mulberry is formed by an entire female inflorescence in which the four tiny floral leaves arranged crosswise and the underlying inflorescence axis become fleshy. The ovaries themselves turn into small, single-seeded drupes whose tiny stones form the hard bits of the fruit; fruit, c. 2.5cm long. The microscopic detail shows the individual fruitlets with their withering stigma remains; fruitlet, 5.3mm wide
Page 108: coastal wattle (*Acacia cyclops,* Leguminosae) – native to south-western Australia; seed surrounded by a bright orange aril to attract birds for dispersal. The peculiar aril is formed by a double layer of the 'umbilical cord' (funicle) which encircles the seed first in one direction and then folds back to surround it once more in the opposite direction; seed, 9mm long (including aril). Nutmeg (Myristica fragrans, Myristicaceae) – native to the Moluccas; the fruit contains a single seed (the nutmeg of commerce) wrapped in a crimson-red, fleshy, lace-like aril that yields the spice called 'mace'. In the wild, the colourful display attracts imperial pigeons (*Ducula* spp.) and hornbills (family Bucerotidae), both capable of swallowing the large seeds; seed, c. 3cm long
Page 109: African mahogany (*Afzelia africana*, Leguminosae) – native to tropical Africa; open fruit revealing the large black seeds with their edible, bright orange appendages, a clear adaptation to attract birds for dispersal; fruit, 17.5cm long. Titoki tree (*Alectryon excelsus*, Sapindaceae) – native to New Zealand; the inconspicuous greenish-brown fruits open with an irregular split from which black seeds wrapped in scarlet, fleshy arils emerge. The speed with which birds discover and collect the seeds is proof of the success of the titoki's ornithochorous strategy; fruits, 8-12mm long
Page 110: snow wood (*Pararchidendron pruinosum*, Leguminosae) – native to Malesia, New Guinea and eastern Australia; the colour scheme suggests the bird-dispersal syndrome but the fruits hold no edible reward for birds. Still a controversial concept, fruit mimicry, i.e. one fruit imitating the appearance of another one, may at least fool some young, inexperienced frugivorous birds into swallowing the hard seeds; fruit, 8-12cm long
Page 111: North Island broom (*Carmichaelia aligera*, Leguminosae) – native to New Zealand; the display of hard, brightly coloured seeds in such a conspicuous way strongly suggests adaptation to bird-dispersal. However, the fact that the fruit holds no edible reward raises the suspicion of deceit; fruit, c. 1cm long. Paternoster pea, crab's eye (*Abrus precatorius,* Fabaceae) – found in all tropical regions; with their red colour the hard, shiny seeds resemble fleshy fruits adapted to bird-dispersal but although beautiful they are poisonous. The seeds of this pantropical climber are very popular with makers of botanical jewellery; seed, diameter 4mm
Page 112: *Floscopa glomerata* (Commelinaceae) – native to Africa; seed, 1.5mm wide
Page 113: Cornflower (*Centaurea cyanus*, Asteraceae) – native to Eurasia and North Africa; the tuft of short, stiff bristles corresponds to the 'parachute' (pappus) in the wind-dispersed fruits of the related dandelion (*Taraxacum officinale*). However, the scale-like pappus segments of the cornflower are neither suitably arranged nor large enough to play any role in wind-dispersal. Instead the scales repeatedly move inwards and outwards with changing humidity, pushing the fruits over the ground for a few centimetres. Movement in the opposite direction is avoided by very short, forward-pointing teeth arranged along the margin of the pappus scales. To achieve further dispersal by ants the cypsela possesses an edible 'oil body' (elaiosome) at the base; fruit, 6mm long
Page 114: Wind-dispersed seeds of larkspurs wearing a dress of helically arranged, papery lamellae: *Consolida orientalis* (origin – southern Europe; seed, diameter 1.8mm). *Delphinium peregrinum* (origin – Mediterranean; seed, diameter 1.2mm)
Page 115: *Delphinium requienii* (origin – southern France, Corsica and Sardinia; seed, 2.6mm long)
Pages 116/7: living rock cactus (*Ariocarpus retusus*, Cactaceae) – native to Mexico; seed; resembling rocks to camouflage their appearance, the c. 8 species of the genus *Ariocarpus* are among the slowest growing cacti, often taking a decade before flowering the first time; seed, 1.5mm long. Detail of seed coat; at high magnification (300x) the convex papillae, each of which represents a single cell of the seed coat, show an intricate pattern of wrinkles, which are the result of a folding of the cuticle, a waxy layer that covers the seed coat
Pages 118/119: wild leek (*Allium ampeloprasum*, Alliaceae) – native to Eurasia and North Africa; the flattened shape of the seeds indicates adaptation to wind-dispersal; seeds 2.9mm long. Detail of seed coat [SEM x500]
Pages 120/121: California fishhook cactus (*Mammillaria dioica*, Cactaceae) – native to California and Mexico; seed 1.1mm long. Detail of seed coat [SEM x500]
Pages 122/123: fringed grass-of-Parnassus (*Parnassia fimbriata,* Parnassiaceae) – native to North America; seed with loose, bag-like seed coat displaying the typical honeycomb-pattern of wind-dispersed balloon seeds. Detail of seed coat (900x)
Page 124: purple foxglove *(Digitalis purpure*a, Plantaginaceae) – native to western Europe and North Africa; seed, 1.3mm long. Detail of seed coat [SEM x500]
Page 125: *Trichodesma africanum* (Boraginaceae) – native to North Africa and the Arabian Peninsula; single-seeded nutlet; 3.9mm long. Surface detail [SEM x90] showing the fringe of tiny hooked spines with which the fruit attaches itself to passing animals; its flat shape may also facilitate wind-dispersal
Pages 126/127: Winter's bark tree (*Drimys winteri*, Winteraceae) – native to Central and South America; the two halves of a longitudinally cut flower bud. The sepals form the outer green 'skin' of the bud, followed on the inside by the larger, folded petals. The petals enclose the centre of the flower bud which harbours the stamens (around the periphery) and the carpels (in the centre). The anthers are partly cut open revealing the pollen grains inside the pollen sacs and the longitudinally split carpel in the centre affords a view of the ovules, which will eventually develop into seeds; diameter 3.9mm
Pages 128/129: love-in-a-mist (*Nigella damascena*, Ranunculaceae) – native to the Mediterranean; the seed of this popular blue-flowering garden plant displays an intriguing surface pattern; seed, 2.6mm long. Detail of seed coat [SEM x180]
Page 130: yellow star-of-Bethlehem (*Ornithogalum dubium*, Hyacinthaceae) – native to South Africa; seed with intricate jigsaw-like surface pattern created by the undulating borders between the individual cells of the seed coat; seed 1.1mm long. Detail of seed coat [SEM x300]
Page 131: pygmy buttercup (*Ranunculus pygmaeus,* Ranunculaceae) – native to northern Europe, the eastern Alps, western Carpathians and North America; shoot with flower and fruit; diameter of flower 4mm. Small-flowered buttercup (*Ranunculus parviflorus*, Ranunculaceae) – native to western Europe and the Mediterranean; one of several nutlets produced by a flower; the hooks on the nutlet's surface indicate adaptation to animal dispersal; 3mm long
Page 132: ragged robin (*Lychnis floscuculi*, Caryophyllaceae) – native to Eurasia; seed; the papillose cells of the seed coat are interlocked like the pieces of a jigsaw as shown by the intricate pattern of undulate lines marking the outlines of the individual cells of the seed coat; seed, 0.9mm long
Page 133: Franklin's sandwort (*Eremogone franklinii*, Caryophyllaceae) – native to North America; seed showing the intricate jigsaw-like surface pattern typical of the pink family; seed, diameter 1.3mm
Page 134/135: Nepal iris (*Iris decora*, Iridaceae) – native to the Himalayas; spheroidal pollen grain with numerous reticulate platelets [SEM x1000]. Flower of an *Iris* cultivar from the 'Pacific Coast' group.

alternation of generations: Unlike the life cycle of animals, a plant's life cycle involves two generations in which mitotic cell divisions take place, the diploid sporophyte (with two sets of chromosomes, one set from of each parent) and the haploid gametophyte (with only half the number of chromosomes). The haploid gametophyte produces the sperm and the egg cells. After its fertilisation by a male sperm, the egg cell becomes a diploid zygote that develops into the sporophyte. When mature, the sporophyte produces haploid spores that give rise to gametophytes again and so on. The hypothetical equivalent scenario in animals would be that both the male sperm and the egg cell first grow into two separate organisms which at some point produce gametes to facilitate fertilisation.
angiosperms (Greek: *angeion* = vessel, small container + *sperma* = seed): division of the seed plants (spermatophytes) that bear ovules and seeds in closed fertile leaves (carpels), in contrast to gymnosperms that have exposed ovules and seeds, borne 'naked' on fertile leaves or cone scales. According to the number of leaves (cotyledons) present in the embryo, two major groups are distinguished, the monocotyledons and the dicotyledons. Angiosperms are commonly referred to as 'flowering plants' even though the reproductive organs of some gymnosperms are also borne in structures that fulfil the definition of a flower.
anemoballism (Greek: *anemos* = wind + *ballistes*, from *ballein* = to throw): form of dispersal in which the diaspores are subject to indirect effects of wind, i.e. wind does not transport the diaspore directly but exerts its influence on the fruit. The fruit (mostly a capsule) is usually exposed on a long flexible stalk that swings in the wind, ejecting the diaspores, e.g. sacred lotus (*Nelumbo nucifera*, Nelumbonaceae), poppy (*Papaver rhoeas*, Papaveraceae).
anemochory (Greek: *anemos* = wind + *chorein* = to disperse): dispersal of fruits and seeds by wind.
anther (Medieval Latin: *anthera* = pollen, derived from Greek: *antheros* = flowery, from *anthos* = flower): the pollen-bearing part of the stamen of the angiosperms. An anther consists of two fertile halves called 'thecae', each bearing two pollen sacs, which usually dehisce with longitudinal slits, valves or pores. The two thecae are connected by a sterile part called the 'connective' which is also the point where the anther is fixed to the filament.
antheridium (plur. *antheridia*) (Latin: small anther; 'anther' referring to the pollen-bearing plant of the angiosperms): male sexual organ of a male or bisexual gametophyte producing and containing the male gamete(s). Antheridia are fully developed in mosses, ferns and fern allies in the broadest sense, but do not occur in seed plants.
anthophytes (Greek: *anthos* = flower + *phyton* = plant): literally 'flowering plants', and often used synonymously with angiosperms. But anthophytes include some gymnosperms, the extinct cycad-like Bennettitales, the closely related *Pentoxylon* and the present-day Gnetales order (comprising the three genera *Ephedra*, *Gnetum*, and *Welwitschia*).
aperture: in pollen grains a preformed opening in the pollen wall through which the pollen tube penetrates.
archegonium (plur. *archegonia*) (New Latin, from Greek: *arkhegonos* = offspring; from *arkhein* = to begin + *gonos* = seed, procreation): often flask-shaped, multi-cellular female sexual organ of a female or bisexual gametophyte producing and containing the female egg cell(s). They are fully developed in mosses, ferns and fern allies, but rudimentary in gymnosperms; in angiosperms true archegonia are absent.
aril (Latin: *arillus* = grape seed): edible seed appendages of various origin in gymnosperms and angiosperms. Arils usually provide an edible reward for animal dispersers.
autochory (Greek: *autos* = self + *chorein* = to disperse): self-dispersal.
ballistic dispersal: dispersal of diaspores through direct or indirect catapult mechanisms, i.e. explosively dehiscent fruits or movement of plant parts by wind (anemoballism) and passing animals, respectively.
berry: a fruit whose fruit wall (pericarp) is entirely fleshy.
caltrop: a structure that consists of four spines that are arranged to point to the four corners of a tetrahedron so that however it falls, it will sit on three of the spines with the fourth one pointing up in the air. Caltrops were first used as a means to slow down pursuers on horseback but later proved effective on pneumatic tyres in the motorised age.
calyx (Greek: *kalyx* = cup): all the sepals of a flower, i.e. the outer whorl of floral leaves in a perianth.
capsule (Latin: *capsula*, diminutive of *capsa* = box, capsule): strictly, a dehiscent fruit developing from an ovary that is composed of two or more joined carpels.
carpel (Modern Latin: *carpellum* = little fruit; originally from Greek: *karpos* = fruit): in angiosperms a fertile leaf that encloses one or more ovules. Carpels are usually sub-divided into an ovule-bearing part (ovary), a style and a stigma. The carpels of a flower can either be separate from each other (e.g. as in a buttercup, *Ranunculus* spp.) or joined (e.g. as in an orange, *Citrus* x *sinensis*, where each fruit segment represents one carpel).
compound fruit: a fruit derived from more than one flower.
conifers (Latin: *conus* = cone + *ferre* = to carry, to bear): group of the gymnosperms generally distinguished by needle- or scale-like leaves and unisexual flowers borne in cones. Well-known examples of conifers are pines, spruces and firs.
corolla (Latin *corolla* = small garland or crown): all the petals of a flower, i.e. the inner whorl of floral leaves in a perianth.
cotyledon (Greek: *kotyle* = hollow object; alluding to the often spoon- or bowl-shape of the seed leaves): the first leaf (in monocotyledons) or pair of leaves (in dicotyledons) of the embryo.
cryptogams (Greek: *kryptos* = hidden + *gamein* = to marry, to copulate): old collective term referring to all plants without recognizable flowers. Cryptogams include algae, fungi (although not really plants), mosses, ferns and fern allies. The Greek meaning 'those who copulate in secret' refers to the absence of flowers as obvious indicators of sexual propagation.
dehiscent fruit: a fruit that opens when ripe to release its seeds into the environment.
diaspore (Greek: *diaspora* = dispersion, dissemination): the smallest unit of seed dispersal in plants. Diaspores can be seeds, fruitlets of compound or schizocarpic fruits, entire fruits or even seedlings (e.g. in mangroves).
dicotyledons (Greek: *di* = two + cotyledon): one of the two major groups of the angiosperms distinguished by the presence of two opposite leaves (cotyledons) in the embryo. Other typical characters of the dicotyledons are reticulate leaf venation, floral organs usually in fours or fives, vascular bundles arranged in a circle, a persistent primary root system developing from embryonic root, and secondary thickening (present in trees and shrubs, usually absent in herbaceous plants). The dicotyledons were long considered a homogeneous entity. Only recently they have been split into two groups, the magnoliids and the eudicots.
drupe: an indehiscent fruit with a fleshy mesocarp and a hardened endocarp that produces one or more stones.
elaiosome (Greek: *elaion* = oil + *soma* = body): literally meaning 'oil body'; a general ecological term referring to edible oily appendages of seeds and other diaspores, usually in the context of ant dispersal.
embryo (Latin: *embryo* = unborn foetus, germ, originally from Greek: *embryon*: *en-* = in + *bryein* = to be full to bursting): in plants the young *sporophyte* developing from the egg cell after fertilization.
embryo sac: the female gametophyte of the angiosperms which develops from a haploid cell (called 'megaspore') formed after a meiotic division of a diploid cell in the ovule. After three mitotic divisions the megaspore produces the female gametophyte/embryo sac, which consists a total of eight nuclei distributed over seven cells: three at the micropylar end (the egg cell and two 'synergids'), three 'antipodal' cells at the chalazal end, and one binucleate 'central cell' in between them.
endocarp (Greek: *endon* = inside + *karpos* = fruit): the innermost layer of the pericarp (fruit wall) forming the hard stone around the seed in drupes.
endosperm (Greek: *endon* = inside + *sperma* = seed): nutritive tissue in seeds.
epicarp (Greek: *epi* = on, upon + *karpos* = fruit): the outermost layer of the fruit wall (pericarp), mostly a soft skin or leathery peel.
epizoochory (Greek: *epi* = on, upon + *zoon* = animal + *chorein* = to disperse): dispersal of diaspores on the surface of the body of an animal. Epizoochorous diaspores adhere to the fleece, coat or feathers of animals or the clothes of man by barbs, hooks or sticky substances.
endozoochory (Greek: *endon* = inside + *zoon* = animal + *chorein* = to disperse): dispersal of the diaspores of a plant by being eaten and carried inside the gut of animals (and man); the usually hard, bad-tasting or poisonous seeds or endocarps discourage mastication and are deposited undamaged with the faeces.
family: one of the main units in the hierarchical system of the taxonomic classification of living organisms. The major classification units are (in descending order) class, order, family, genus and species.
filament (Latin: *filum* = thread, string): the stalk of a stamen.
flowering plants: the meaning is regionally different depending on the definition of 'flower'. In continental Europe the term is considered to comprise both gymnosperms and angiosperms, in Anglo-America and the UK it is applied only to angiosperms. In a strict scientific sense, 'flowering plants' are explained under 'anthophytes'.
fruit: any self-contained seed-bearing structure, including domesticated fruits that have been bred to be seedless.
fruit wall: part of the fruit that is derived from the wall of the ovary, also called 'pericarp'.
fruitlet: a separate dispersal unit of a fruit that may be (1) a carpel or half-carpel of a mature schizocarpic fruit, (2) a single carpel of mature multiple fruit, or (3) a mature (mono- or multicarpellate) ovary of a compound fruit.
funicle (Latin: *funiculus* = slender rope): the stalk by which an ovule or seed is connected to the placenta in the ovary. The funiculus acts like an 'umbilical cord', supplying the developing ovule and seed with water and nutrients from the parent plant.
gamete (Greek: *gametes* = spouse): haploid male or female sex cells. Male and female gametes fuse upon copulation. In contrast to spores, gametes can only give rise to a new individual or generation after they fused with a gamete of the opposite sex.
gametophyte (Greek: *gametes* = spouse + *phyton* = plant): the haploid generation in a plant's life cycle producing gametes. Examples are the prothallus of the ferns or the germinated pollen grain of the seed plants.
Gnetales: heterogeneous group of gymnosperms comprising just three families with three genera (*Gnetum*, *Ephedra*, *Welwitschia*) and all together 95 species.
gymnosperms (Greek: *gymnos* = naked + *sperma* = seed): non homogeneous group of seed plants bearing their ovules on open fertile leaves (or ovuliferous scales in conifers) and not in closed carpels as is the case in angiosperms. Gymnosperms comprise three distantly related groups: conifers (8 families, 69 genera, 630 species), cycads (3 families, 11 genera, 292 species) and Gnetales (3 families, 3 genera, 95 species).
gynoecium (Greek: *gyne* = woman + *oikos* = house): all the carpels of a flower, irrespective of whether they are joined or separate.
hydrochory (Greek: *hydor* = water + *chorein* = to disperse): dispersal of plant diaspores by water.

indehiscent fruit: a fruit that remains closed even when ripe.
inflorescence: part of a plant which bears a group of flowers; inflorescences can be a loose group of flowers (as in lilies) or highly condensed and differentiated structures resembling an individual flower, such as the capitulae (flower heads) in the sunflower family (Asteraceae).
infructescence: the flowers of an inflorescence at the fruiting stage.
mesocarp (Greek: *mesos* = middle + *karpos* = fruit): the fleshy middle layer of the fruit wall (pericarp).
micropyle (Greek: *mikros* = small + *pyle* = gate): the opening at the apex of the ovule that acts as a passage for the pollen tube on its way to the egg cell.
multiple fruit: a fruit that develops from a gynoecium of two or more separate carpels; each carpel develops into a fruitlet, e.g. a raspberry.
myrmecochory (Greek: *myrmex* = ant + *chorein* = to disperse): dispersal of seeds and other diaspores by ants.
nectar guides: coloured patterns of lines, speckles or larger spots in flowers that guide pollinators to the nectar and pollen. Nectar guides may be visible to the human eye or invisible if they are based on ultraviolet reflection (bees and most other insects are able to see ultraviolet light).
nectaries: glands secreting nectar to attract pollinators. Nectaries are usually situated in the base of the flower or spur (e.g. *Aquilegia* spp.).
nut: a dry, indehiscent and usually single-seeded fruit in which the pericarp is contiguous to the seed.
nutlet: diminutive of 'nut', referring to an individual nut-like fruitlet of a schizocarpic fruit or a fruit that develops from a gynoecium of two or more separate carpels.
ornithochory (Greek: *ornis* = bird + *chorein* = to disperse): dispersal of fruits and seeds by birds.
ovary (New Latin: *ovarium* = a place or device containing eggs, from Latin: *ovum* = egg): the enlarged, usually lower portion of a pistil containing the ovules.
ovule (New Latin: *ovulum* = small egg): the female sexual organ of the seed plants that, after fertilization of its egg cell, develops into the seed.
perianth (Greek: *peri* = around; *anthos* = flower): the floral envelope that is clearly differentiated into calyx (outer perianth whorl) and corolla (inner perianth whorl).
pericarp (New Latin: *pericarpum*, from Greek: *peri* = around + *karpos* = fruit): the wall of the ovary at the fruiting stage. The pericarp can be homogeneous (as in berries) or divided into three layers (as in drupes) called epicarp, mesocarp, and endocarp.
petal (New Latin: *petalum*, from Greek *petalon* = leaf): in flowers where the outer whorl of the perianth is different from the inner whorl, the elements of the inner whorl of the floral envelope are addressed as petals. The petals forms the often brightly coloured, showy corolla of a flower.
pistil (Latin: *pistillum* = pestle; alluding to the shape): an individual ovary with one or more styles and stigmas, composed of one or more carpels. The term was introduced in 1700 by the French botanist Joseph Pitton de Tournefort. Nowadays, because of its unclear meaning, scientists omit its usage and replace it with gynoecium.
placenta (Modern Latin: *placenta* = flat cake, originally from Greek *plakoenta*, accusative of *plakoeis* = flat, related to *plax* = anything flat): a region within the ovary where the ovules are formed and remain attached (usually via a funiculus) to the parent plant until the seeds are mature. In botany the term was adopted from the similar structure to which the embryo is attached in animals and humans.
pollen (Latin for *fine flour*): the microspores of the seed plants, able to germinate on the stigma (in angiosperms) or in the pollen chamber of the ovules (in gymnosperms). The germinated pollen grain with its pollen tube represents a very small, strongly simplified gametophyte.
pollen chamber: a chamber at the apex of the ovules of many gymnosperms where the pollen grains end up and germinate.
pollen sac: the container in which angiosperms produce their pollen; it is homologous to the sporangium of ferns. One anther typically bears four pollen sacs.
pollen tube: tube-like structure formed by the germinating pollen grain. In cycads and *Ginkgo* the pollen tube releases motile sperm directly into the pollen chamber from where they swim to the archegonia. In conifers and angiosperms the pollen tube delivers the naked and non-motile sperm nuclei straight to the egg cells.
pollenkitt: a sticky substance composed mainly of saturated and unsaturated lipids, carotenoids, proteins and carboxylated polysaccharides. It is found in all angiosperms so far studied, but seems to be absent from mosses (bryophytes), ferns (pteridophytes) and gymnosperms. It has various functions: containing the proteins inside the pollen wall; keeping pollen grains in or near the anthers until collection by pollinating animals; holding pollen grains in clumps so that they reach the stigma together in larger pollen 'parcels'; allowing adhesion to insect bodies, birds beaks, etc.; protecting the cytoplasm of pollen grains from solar radiation; preventing excessive loss of water from the cytoplasm; determining the colour of pollen; attracting pollinators with oily and perfumed components.
pollination syndrome: a suite of flower characteristics that have evolved as a result of adaptation to a certain mode of pollen transfer, such as pollination by wind, water and animals.
pollinium (plur. *pollinia*): a structure in which the individual pollen grains remain massed together, to be transported as a single unit during pollination.
polyads (Greek: *poly* = many): groups of pollen grains which remain attached at maturity and are dispersed as a unit. Polyads are usually groups of pollen in multiples of four grains.
prothallus (plur. *prothalli*) (Greek: *pro* = before, in front + *thallos* = shoot): a small haploid (male, female or hermaphrodite) gametophyte. Prothalli are well developed in algae, mosses, ferns and fern allies and some gymnosperms. A prothallus develops from a haploid spore and produces either antheridia or archegonia or both. In angiosperms both male and female gametophytes are highly reduced (without formation of antheridia and archegonia) with the pollen tube and embryo sac representing the male and female gametophytes.
samara (Latin name for the fruit of the elm, *Ulmus*): a winged nut.
schizocarpic fruit (New Latin, from Greek: *skhizo-*, from *skhizein* = to split + *karpos* = fruit): one in which the carpels are partially or completely joined at the time of pollination but separate from each other at maturity, each part functioning as a diaspore.
seed: the organ of the seed plants that encloses the embryo together with a nutritious tissue within a protective seed coat. Seeds develop from ovules, the defining organ of the seed plants.
seed plants: plants that produce seeds, see *spermatophytes*.
sepal (New Latin: *sepalum*, an invented word, perhaps a blend of Latin: *petalum* and Greek: *skepe* = cover, blanket): in flowers where the outer whorl of the perianth is different from the inner whorl, the elements of the outer whorl are addressed as sepals. The summary of the sepals forms the generally inconspicuous green calyx of a flower.
sorus (plur. *sori*): a cluster of sporangia on the underside of fern fronds.
sperm nucleus: the extremely reduced, non-motile male gamete of conifers and angiosperms.
spermatophytes (Greek: *spermatos* = seed + *phyton* = plant): seed-producing plants. The spermatophytes comprise two major groups, the gymnosperms and the angiosperms.
sporangium (Greek: *sporos* = germ, spore + *angeion* = vessel, container): container with an outer cellular wall and a core of cells which give rise to spores.
spore: a cell serving asexual reproduction.
sporophyte (*sporos* = germ, spore + *phyton* = plant): literally 'the plant which produces the spores'; the diploid generation in the life cycle of plants which produces asexual, haploid spores that give rise to haploid gametophytes.
stamen (Latin: *stamen* = thread): the pollen-producing organ of the angiosperms, consisting of the sterile filament that carries the fertile anther at the apex; each anther has four pollen sacs containing the pollen grains.
stigma (Greek = spot, scar): specialised area of the gynoecium of the angiosperms that is able to receive pollen grains and facilitate their germination; the stigma is usually elevated above the ovary by a style.
style (Greek: *stylos* = column, pillar): in angiosperms the narrow, elongated part of a carpel or pistil connecting stigma and ovary through which the pollen tubes grow down into the ovary.
tetrad (Greek: *tetra* = four): a general term for a group of four united pollen grains or spores, either as a dispersal unit or as a developmental stage.
zoochory (Greek: *zoon* = animal + *chorein* = to disperse): dispersal of fruits and seeds by animals.
zygote (Greek: *zygotos* = joined together): a fertilised (diploid) egg cell.

PICTURE CREDITS

p 28 bottom right: © Paula Rudall, Jodrell Laboratory, Royal Botanic Gardens, Kew; p 28 left, top & bottom: © Hannah Banks, Jodrell Laboratory, Royal Botanic Gardens, Kew; p 38, 84: © Merlin Tuttle, Bat Conservation International; p 44 bottom, left & right: © Dr. Klaus Schmitt, Weinheim, http://photographyoftheinvisibleworld.blogspot.com/; p 48: © Paul Garfinkel, River Road Photography, Jacksonville Florida, USA, www.riverroadphoto.net; p 49 top right: © Prof. Dr. Lutz Thilo Wasserthal, Institut für Zoologie, Friedrich-Alexander-Universität, Erlangen, Germany; p 54: © Mark G.; p 56: © Timothy Motley; p 57, 108: © Dr. Gerald Carr; p 58: ZO-847 © Jiri Lochman / Lochman Transparencies; p 109 top right: © Dr. Trevor James, New Zealand.

BIBLIOGRAPHY

Armstrong, W.P. A non-profit natural history textbook dedicated to little-known facts and trivia about natural history subjects. www.waynesword.com
Bell, A.D. (1991) *Plant form – an illustrated guide to flowering plant morphology*, Oxford University Press, UK
Fenner, M. & Thompson, K. (2005) *The ecology of seeds*, Cambridge University Press, Cambridge, UK
Gunn, C.R. & Dennis, J.V. (1999) *World guide to tropical drift seeds and fruits* (reprint of the 1976 edition), Krieger Publishing Company, Malabar, Florida, USA
Heywood, V.H., Brummit, R.K., Culham, A. & Seberg, O. (2007) *Flowering Plant Families of the World*, Royal Botanic Gardens, Kew, London, UK.
Janick, J. & Paull, R.E. (eds.) (2008) The encyclopedia of fruit and nuts, CABI Publishing, UK
Janzen, D.H. (1984) Dispersal of small seeds by big herbivores: foliage is the fruit, *The American Naturalist* 123: 338-353
Judd, W.S., Campbell, S., Kellogg, E.A., Stevens, P.F. & M.J. Donoghue (2002) *Plant Systematics - a phylogenetic approach*, Sinauer Associates, Inc., Sunderland, MA, USA
Kesseler, R. & Harley, M. (2009) *Pollen – The Hidden Sexuality of Flowers*, 3rd edition, Papadakis Publisher, London, UK
Kesseler, R. & Stuppy, W. (2009) *Seeds – Time Capsules of Life*, 2nd edition, Papadakis Publisher, London, UK
Loewer, P. (2005) *Seeds – the definitive guide to growing, history and lore*, Timber Press, Portland, Cambridge, USA
Mabberley, D.J. (2008) *Mabberley's Plant-Book*, 3rd edition, Cambridge University Press, UK
Mauseth, J.D. (2003) *Botany - an introduction to plant biology*, 3rd edition, Jones and Bartlett Publishers Inc., Boston, USA
Pijl, L. van der (1982) *Principles of dispersal in higher plants*, 3rd edition, Springer, Berlin, Heidelberg, New York
Raven, P.H., Evert, R.F. & Eichhorn, S.E. (1999) *Biology of plants*, W.H. Freeman, New York, USA
Ridley, H.N. (1930) *Dispersal of plants throughout the world*, L. Reeve & Co., Ashford, Kent, UK
Spjut, R.W. (1994) A systematic treatment of fruit types, *Memoirs of the New York Botanical Garden* 70: 1-182
Stuppy, W. & Kesseler, R. (2008) *Fruit – Edible, Inedible, Incredible*, Papadakis Publisher, London, UK
Ulbrich, E. (1928) *Biologie der Früchte und Samen (Karpobiologie)*, Springer, Berlin, Heidelberg, New York

INDEX OF PLANTS ILLUSTRATED

Family	Species	Common name	Distribution
Euphorbiaceae	***Euphorbia helioscopia*** L.	Sun Spurge *96*	Eurasia and North Africa
Euphorbiaceae	***Euphorbia peplus*** L.	Petty Spurge *96, 97*	Eurasia
Euphorbiaceae	***Euphorbia punicea*** Sw.	Jamaican Poinsettia *10*	Jamaica
Gentianaceae	***Eustoma grandiflorum*** (Raf.) Shinners	Lisianthus, Prairie Gentian *29*	America and the Caribbean
Moraceae	***Ficus carica*** L.	Common Fig *101*	Ancient cultigen, esp. Mediterranean, originally probably from South-West Asia
Moraceae	***Ficus villosa*** Blume	Villous Fig, Shaggy Fig *85*	Tropical Asia
Commelinaceae	***Floscopa glomerata*** (Willd. ex Schult. & Schult. f.) Hassk.	no common name *112*	Africa
Rosaceae	***Fragaria x ananassa*** (Weston) Decne & Naudin	Garden Strawberry *60, 99*	Only in cultivation
Oleaceae	***Fraxinus ornus*** L.	Manna Ash *21*	Eurasia
Asteraceae	***Galinsoga brachystephana*** Regel	no common name *66*	Central and South America
Rubiaceae	***Galium aparine*** L.	Stickywilly, Goosegrass, Sticky Bobs *87*	Eurasia, America
Clusiaceae	***Garcinia arenicola*** (Jum. & H. Perrier) P. Sweeney & Z.S. Rogers	no common name *25*	Madagascar
Asteraceae	***Gazania krebsiana*** Less.	Terracotta Gazania *35*	Namaqualand (South Africa)
Ginkgoaceae	***Ginkgo biloba*** L.	Ginkgo, Maidenhair Tree *24*	A relic in eastern China
Boraginaceae	***Hackelia deflexa*** (Opiz) **var. *americana*** (A. Gray) Fernald and I.M. Johnst.	American Stickseed, Nodding Stickseed *86*	North America
Pedaliaceae	***Harpagophytum procumbens*** DC. ex Meisn.	Devil's Claw, Grapple Plant *93*	Southern Africa, Madagascar
Ranunculaceae	***Helleborus orientalis*** Lam.	Lenten Rose *30, 41*	Greece, Turkey
Lamiaceae	***Hemizygia transvaalensis*** Ashby	Mpumalanga Sagebush *32*	South Africa
Malvaceae	***Heritiera littoralis*** Aiton	Looking-glass Mangrove *79, 80*	Old World Tropics
Malvaceae	***Hibiscus mutabilis*** L.	Confederate Rose, Cotton Rosemallow, Dixie Rosemallow *7*	Native to China and Japan, naturalised in the southern USA
Leguminosae	***Hippocrepis unisiliquosa*** L.	Horse-shoe Vetch *12*	Eurasia and Africa
Apocynaceae	***Hueruia hislopii*** Turrill	no common name *53*	Africa
Euphorbiaceae	***Hura crepitans*** L.	Sandbox Tree *24*	South America and the Caribbean
Cactaceae	***Hylocereus undatus*** (Haw.) Britton & Rose	Dragon Fruit, Pitahaya *100*	Tropical America
Rubiaceae	***Hymenodictyon floribundum*** (Hochst. & Steud.) B.L. Rob	Firebush *67*	Native to Africa
Balsaminaceae	***Impatiens glandulifera*** Royle	Himalayan Balsam *83*	Himalayas
Balsaminaceae	***Impatiens tinctoria*** A.Rich.	no common name *50*	Africa
Iridaceae	***Iris decora*** Wall.	Nepal Iris *135*	Himalayas (Pakistan to South-West China)
Bignoniaceae	***Kigelia africana*** (Lam.) Benth.	Sausage Tree *56*	Tropical Africa
Krameriaceae	***Krameria erecta*** Willd. ex Schult.	Littleleaf Rhatany, Pima Rhatany *88*	Southern USA and Northern Mexico
Orobanchaceae	***Lamourouxia viscosa*** Kunth	no common name *74*	Mexico
Lamiaceae	***Leonurus cardiaca*** L.	Motherwort *30*	Central Asia
Asteraceae	***Leucochrysum molle*** (DC.) Paul G. Wilson	Hoary Sunray, Golden Paper Daisy *72, 73*	Australia
Liliaceae	***Lilium speciosum*** Thunb. **var. *clivorum*** Abe et Tamura	Japanese Lily *26*	Japan
Sapindaceae	***Litchi chinensis*** Sonn. **subsp. *chinensis***	Lychee *101*	South China
Loasaceae	***Loasa chilensis*** (Gay) Urb. & Gilg	no common name *74, 75*	Chile
Arecaceae	***Lodoicea maldivica*** Pers.	Seychelles Nut, Coco-de-Mer *82*	Seychelles (Curieuse and Praslin only)
Caryophyllaceae	***Lychnis flos-cuculi*** L.	Ragged Robin *132*	Eurasia
Malvaceae	***Malva sylvestris*** L.	Common Mallow *45*	Eurasia
Cactaceae	***Mammillaria dioica*** K. Brandegee	California Fishhook Cactus *120, 121*	Mexico
Anacardiaceae	***Mangifera indica*** L.	Mango *101*	Cultigen, probable origin somewhere between India and the Malay peninsula
Solanaceae	***Markea neurantha*** Hemsl.	no common name *38*	Mexico to Panama, widely cultivated in the tropics
Leguminosae	***Medicago polymorpha*** L.	Burclover, Toothed medick *91*	Eurasia, North Africa
Cactaceae	***Melocactus zehntneri*** (Britton & Rose Luetzelb)	Turk's-cap Cactus *25*	Brazil
Convolvulaceae	***Merremia discoidesperma*** (Donn. Sm.) O'Donnell	Mary's Bean, Crucifixion Bean *80*	Central America
Leguminosae	***Mora megistosperma*** (Pittier) Britton & Rose	Nato Mangrove *82*	Tropical America
Morinaceae	***Morina longifolia*** Wall. ex DC.	Whorlflower *7*	Native to Himalayas (from Kashmir to Bhutan)
Moraceae	***Morus nigra*** L.	Black Mulberry *106, 107*	Cultivated since antiquity; originally probably from China
Leguminosae	***Mucuna urens*** (L.) Medik.	Hamburger Bean, Horse-eye Bean *80, 81*	Central and South America
Myristicaceae	***Myristica fragrans*** Houtt.	Nutmeg *108*	Indonesia (Moluccas)
Plantaginaceae	***Nemesia versicolor*** E. Mey. ex Benth.	Leeubekkie (Afrikaans) *66*	South Africa, Eastern Cape
Amaryllidaceae	***Nerine bowdenii*** S. Watson	no common name *26*	South Africa
Ranunculaceae	***Nigella damascena*** L.	Love-in-a-mist, Devil-in-the-bush (US) *128, 129*	Mediterranean
Menyanthaceae	***Nymphoides peltata*** (S.G. Gmel.) Kuntze	Yellow Floatingheart *79*	Eurasia
Arecaceae	***Nypa fruticans*** Wurmb	Nipa Palm, Mangrove Palm *78*	Southern Asia to Northern Australia, naturalised in West Africa and Panamá.
Orchidaceae	***Ophrys apifera*** Huds.	Bee Orchid *43*	Native to Central and Southern Europe (including Britain) and North Africa
Apocynaceae	***Orbea lutea*** (N.E. Br.) Bruyns	no common name *42*	Southern Africa
Hyacinthaceae	***Ornithogalum dubium*** Houtt.	Yellow Star-of-Bethlehem *130*	South Africa, Western Cape
Orobanchaceae	***Orobanche*** sp.	Broomrape *71*	Collected in Greece
Orobanchaceae	***Orthocarpus luteus*** Nutt.	Yellow Owl's Clover *25*	North America
Melastomataceae	***Osbeckia crinita*** Benth.	no common name *20*	Native to Eastern Asia
Papaveraceae	***Papaver rhoeas*** L.	Corn Poppy, Field Poppy *21, 77*	Eurasia and North Africa
Leguminosae	***Pararchidendron pruinosum*** (Benth.) I.C. Nielsen	Snow Wood, Monkey's Earrings *110*	Eastern Australia, New Guinea and Malesia
Parnassiaceae	***Parnassia fimbriata*** K.D. Koenig **var. *fimbriata***	Fringed Grass-of-Parnassus *122, 123*	North America
Passifloraceae	***Passiflora edulis*** Sims forma ***edulis***	Purple Passionfruit *100*	South America
Paulowniaceae	***Paulownia tomentosa*** (Thunb.) Steud.	Princess Tree *69*	Native to China, widely cultivated
Malvaceae	***Pavonia spinifex*** (L.) Cav.	Gingerbush *20*	Tropical America
Proteaceae	***Persoonia mollis*** R.Br.	no common name *32*	Australia
Picrodendraceae	***Petalostigma pubescens*** Domin	Quinine Bush *95*	Malesia and Northern and Eastern Australia
Phytolaccaceae	***Phytolacca acinosa*** Roxb.	Indian Pokeweed *43*	East Asia (China to India)
Pinaceae	***Pinus radiata*** D. Don	Monterey Pine *37*	California
Poaceae	***Poa trivialis*** L.	Rough Meadow Grass *36, 37*	Native to Eurasia and North Africa
Polygalaceae	***Polygala arenaria*** Oliv.	Sand Milkwort *97*	Tropical Africa
Martyniaceae	***Proboscidea altheifolia*** (Benth.) Decne.	Golden Devil's Claw, Desert Unicorn-plant, Yuca de Caballo *92*	Southern USA and Mexico
Rosaceae	***Prunus dulcis*** (Mill.) D.A. Webb (syn. ***Prunus amygdalus*** Batsch)	Almond *18, 19*	Western Asia (Levant)
Rosaceae	***Prunus persica*** (L.) Batsch **var. *persica***	Peach *101, 104, 105*	Originally from China, widely cultivated
Lythraceae	***Punica granatum*** L.	Pomegranate *100*	Native to Asia (Middle East to Himalaya)
Rosaceae	***Pyrus pyrifolia*** (Burm. f.) Nakai	Chinese Pear, Japanese Pear, Nashi Pear, Apple Pear *101*	East Asia (China, Laos, Vietnam)
Ranunculaceae	***Ranunculus acris*** L.	Meadow Buttercup *21*	Eurasia
Ranunculaceae	***Ranunculus parviflorus*** L.	Small-flowered Buttercup *131*	Western Europe and Mediterranean, naturalised elsewhere in temperate areas
Ranunculaceae	***Ranunculus pygmaeus*** Wahlenb.	Pygmy Buttercup, Dwarf Buttercup *131*	Northern Europe, eastern Alps, western Carpathians, North America
Ericaceae	***Rhododendron*** cv. 'Naomi Glow'	Rhododendron *40*	Cultivated variety
Rosaceae	***Rubus phoenicolasius*** Maxim.	Japanese Wineberry, Wineberry *98*	Northern China, Korea, Japan; cultivated in Europe and North America
Asteraceae	***Rudbeckia hirta*** 'Prairie Sun'	Rudbeckia *44*	Cultivated variety
Salicaceae	***Salix caprea*** L.	Pussy Willow *14*	Eurasia
Lamiaceae	***Salvia dorisiana*** Standl.	Fruit-scented Sage *32*	Honduras

Dipsacaceae	***Scabiosa crenata*** Cyr.	No common name *69*	Central and eastern Mediterranean
Selaginellaceae	***Selaginella lepidophylla*** (Hook. & Grev.) Spring	False Rose of Jericho, Flower of Stone *23*	Chihuahuan Desert (USA, Mexico)
Caryophyllaceae	***Silene dioica*** (L.) Clairv.	Red Campion *47, 76, 77*	Europe
Caryophyllaceae	***Silene gallica*** L.	Small-flowered Catchfly *62*	Eurasia and North Africa
Caryophyllaceae	***Silene maritima*** With.	Sea Campion *62*	Europe
Solanaceae	***Solanum betaceum*** Cav. (syn. ***Cyphomandra betacea*** (Cav.) Sendtn.)	Tree-tomato, Tamarillo *100*	South America, widely cultivated in the tropics
Caryophyllaceae	***Spergularia media*** (L.) C. Presl.	Greater Sea-spurrey, Greater Sand-spurrey *66*	Eurasia and North Africa
Caryophyllaceae	***Spergularia rupicola*** Lebel ex Le Jolis	Rock Sea-spurrey *70*	Europe
Orchidaceae	***Stanhopea tigrina*** Bateman ex Lindl.	Tiger-like Stanhopea *71*	Eastern Mexico to Brazil
Caryophyllaceae	***Stellaria holostea*** L.	Greater Stitchwort *33*	Europe
Caryophyllaceae	***Stellaria pungens*** Brogn.	Prickly Starwort *63*	Australia
Papaveraceae	***Stylophorum diphyllum*** (Michx.) Nutt.	Poppywort, Celandine Poppy *97*	Eastern United States
Leguminosae	***Swainsona formosa*** J. Thomps.	Sturt's Desert Pea *13*	Australia
Boraginaceae	***Symphytum officinale*** L.	Comfrey *28*	Europe
Proteaceae	***Telopea speciosissima*** (Sm.) R.Br.	Waratah *55*	Australia (New South Wales)
Gyrostemonaceae	***Tersonia cyathiflora*** (Fenzl) A.S. George ex J.W. Green	Button Creeper *96*	Australia, Western Australia
Saxifragaceae	***Tolmiea menziesii*** (Hook.) Torr. & A. Gray	Piggyback Plant, Youth-on-age, Mother-of-thousands *70*	USA, Oregon
Araliaceae	***Trachymene ceratocarpa*** (W. Fitzg.) Keighery & Rye	Creeping Carrot *91*	Australia
Zygophyllaceae	***Tribulus terrestris*** L.	Puncture Vine, Caltrop, Devil's Thorn *93*	Native to the Old World
Boraginaceae	***Trichodesma africanum*** (L.) Lehm.	No common name *125*	North Africa and from the Arabian Peninsular to Iran
Cucurbitaceae	***Trichosanthes cucumerina*** L.	Snake Gourd *51*	Asia
Pedaliaceae	***Uncarina*** sp.	no common name *92*	Collected in Madagascar
Valerianaceae	***Valerianella coronata*** (L.) DC.	no common name (literally 'crowned lambs lettuce') *69*	Central and Southern Europe, North Africa, Turkey, South-West and Central Asia, Afghanistan, Pakistan

ACKNOWLEDGMENTS

Many people have, either directly, or indirectly, contributed to the amazing wealth of plant specimens, knowledge and ideas that have provided the material for this book. Although we cannot list all the scientists whose painstaking observations and publications over decades have revealed so many fascinating facts about the life of plants, or all the people who discovered, collected or grew the plants that are the source of the images shown here, we can and should give special mention to the following colleagues and friends:

We thank our publisher, the late Andreas Papadakis, for the freedom, inspiration and generous support he gave us throughout the preparation of our previous three books; we also want to thank his daughter Alexandra for using her creative vision in realising this spectacular new book, which fuses text with images in such a visually powerful way.

We are immensely appreciative of the unique opportunity afforded to us by the Royal Botanic Gardens, Kew, which has enabled the development of our latest book; we are especially grateful to the present and previous directors, Stephen Hopper and Sir Peter Crane, Paul Smith, Head of the Seed Conservation Department (SCD) of the Royal Botanic Gardens, Kew, and John Dickie (SCD) for their continuing support of Wolfgang Stuppy's work .We are deeply indebted to all members of the Seed Conservation Department, and the many partners of the Millennium Seed Bank Project all over the world who have contributed towards the outstanding collection of seeds and fruits that have provided much of the material we used for the images in this book.

The Millennium Seed Bank Project is funded by the UK Millennium Commission and the Wellcome Trust. The Royal Botanic Gardens, Kew receives an annual grant in aid from the UK Department of Environment, Food and Rural Affairs.

The University of the Arts London and Central Saint Martins College of Art & Design have played a vital role in continuously supporting Rob Kesseler's work since the inception of the project in 1999. In particular we would like to thank Jane Rapley OBE (Head of College), Jonathan Barratt (Dean of Graphic and Industrial Design), Kathryn Hearn (Course Director, BA Ceramic Design), and the many colleagues from whom the work has received such positive acknowledgment. NESTA (The National Endowment for Science Technology and the Arts) provided timely support for his Fellowship that led to the first publication (on Pollen) under the enthusiastic guidance of Alex Barclay.

We are indebted to Stephen Blackmore (Regius Keeper, Edinburgh Botanic Garden) for critical reading of early versions of *Pollen, the Hidden Sexuality of Flowers*, and to Richard Bateman, Paula Rudall and Richard Spjut for their thorough reviews of the manuscripts of *Seeds – Time Capsules of Life* and *Fruit – Edible, Inedible, Incredible*.

At Kew, we would like to thank members of the Legume and Palm Sections in the Herbarium for

Paula Rudall, Head of the Micromorphology Section, for allowing us to use the scanning electron microscope (SEM) and ancillary equipment; we also thank Chrissie Prychid and Hannah Banks for technical support. At the SCD, we are grateful for the kind support of the members of the Curation Section.

There are many present, and former colleagues at the Royal Botanic Gardens, Kew who we wish to thank for their generously offered expertise when we needed answers to difficult questions, or support in providing us with important materials including photographs, in particular: John Adams (SCD), Steve Alton (SCD), Bill Baker (Herbarium), Mike Bennett (former Keeper of the Jodrell Laboratory), David Cooke (HPE), Tom Cope (Herbarium), Sir Peter Crane (former Director of the Royal Botanic Gardens, Kew, now Dean, School of Forestry and Environmental Studies, Yale University), Matthew Daws (formerly SCD, now Rio Tinto), John Dickie (SCD), John Dransfield (Herbarium), Laura Giuffrida (HPE), Anne Griffin (Library), Phil Griffiths (Horticulture & Public Education – HPE), Tony Hall (retired, formerly HPE, now Kew Research Associate), Chris Haysom (HPE), Steve Hopper (Director of the Royal Botanic Gardens, Kew), Kathy King (HPE), Tony Kirkham (HPE), Ilse Kranner (SCD), Gwilym Lewis (Herbarium), Mike Marsh (HPE), Mark Nesbitt (Centre for Economic Botany), Simon Owens (former Keeper of the Herbarium), Grace Prendergast (Micropropagation), Hugh Pritchard (SCD), Chrissie Prychid (Jodrell Laboratory), Brian Schrire (Herbarium), Wesley Shaw (HPE), Nigel Taylor (Curator, HPE), Janet Terry (SCD), James Wood (now Royal Tasmanian Botanical Gardens), Elly Vaes (SCD), Suzy Wood (SCD) and Daniela Zappi (Herbarium).

For assistance in the field we would like to personally thank Sarah Ashmore, Phillip Boyle, Andrew Crawford, Richard Johnstone, Andrew Orme, Andrew Pritchard and Tony Tyson-Donnelly in Australia; Ismael Calzada and Ulises Guzmán in Mexico; and Michael Eason and Patricia Manning in Texas (USA). In South Africa, we would like to thank Ernst van Jaarsveld and Anthony Hitchcock (Kirstenbosch Botanical Garden, Cape Town) for their time, hospitality, and permission to photograph plants in their collections; in Australia, the staff of Kings Park and Botanic Garden, Perth; Geelong Botanic Gardens; Brisbane Botanic Gardens at Mt Coot-tha; the Royal Botanic Gardens, Melbourne; the Royal Botanic Gardens, Sydney; and Mount Annan Botanic Garden, New South Wales for their hospitality and permission to photograph plants in their collections. In New Zealand WS would like to thank his friend and colleague Trevor James for his hospitality and company in the field while visiting his country and for photographing the fruits of the titoki tree (*Alectryon excelsus*) for us.

At Papadakis, our thanks to Sheila de Vallée and Sarah Roberts for editing the text and Naomi Doerge for her help with the production of this book.